Maternal Influences on Fetal Neurodevelopment

Andrew W. Zimmerman • Susan L. Connors
Editors

Maternal Influences on Fetal Neurodevelopment

Clinical and Research Aspects

Springer

Editors
Andrew W. Zimmerman
Director of Medical Research
Center for Autism and Related Disorders
Kennedy Krieger Institute
Associate Professor of Neurology
Psychiatry and Pediatrics
Johns Hopkins University School of Medicine
Baltimore, MD
USA
zimmerman@kennedykrieger.org

Susan L. Connors
Internal Medicine and Pediatrics
Lurie Family Autism Center / LADDERS
Clinic
Massachusetts General Hospital for Children
Instructor in Medicine
Harvard Medical School
Boston, MA
USA
slconnors@partners.org

ISBN 978-1-4899-8161-5 ISBN 978-1-60327-921-5 (eBook)
DOI 10.1007/978-1-60327-921-5
Springer New York Dordrecht Heidelberg London

Printed on acid-free paper

Springer is part of Springer Science+Business Media (www.springer.com)

Preface

Novel Approaches into the Origins of Neurodevelopmental Disorders: The Fetal Physiology Foundation

Over the past two decades, *autism*, a neurodevelopmental disorder that is defined by behavior and was once believed to be rare, became recognized in increasing numbers of children and recently received distinction as an "epidemic" [1]. While numbers of affected children have steadily increased, our knowledge is still insufficient to explain autism's diverse causes and broad range of presentations. Despite remarkable progress in research, available medical diagnostic testing applies only to a small minority of affected children. Thus, scientifically based explanations with which physicians can diagnose and treat the majority of children with autism and advise their parents are quite limited.

Our society and scientific community were unprepared for the rise in autism, which explains our present inability to understand most of its causes. Researchers in neurodevelopmental disorders have long been aware of other disorders that, despite extensive efforts, have not yielded clear genetic or environmental origins, and autism has become symbolic of the need for new approaches to research into these complex conditions. Although autism has captured our attention in recent years, the prevalence of other neurodevelopmental disorders such as attention deficit hyperactivity disorder (ADHD) and bipolar disorder, among others, also has been increasing [2–4]. Several of these conditions share some symptoms with autism, and ADHD, bipolar disorder, OCD, Tourette syndrome and schizophrenia occur more frequently than expected in the extended families of children diagnosed with autism. Similar to autism, none of these disorders is likely to have singular genetic or environmental causes, even though both genes and environmental factors have been implicated in their origins [5].

Further evidence for relationships among these neurodevelopmental disorders can be observed in their overlapping symptoms (Fig. 1). Hyperactivity is present in ADHD and frequently, in autism. Problems with mood regulation are seen in bipolar disorder, ADHD and autism. Thought disorder occurs in schizophrenia and often in bipolar disorder, and difficulty relating to others is common in autism, and may be seen in individuals with ADHD and bipolar disorder. Clinicians have long

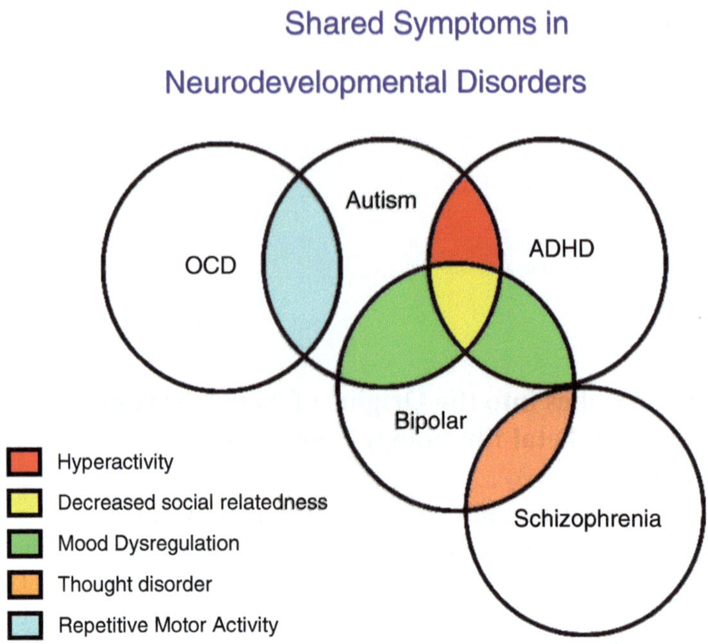

Fig. 1 Several neurodevelopmental disorders have symptoms in common that may result from shared cellular mechanisms during fetal life which may lead to abnormal brain development

observed that family members of affected individuals frequently display traits of these disorders, although they are usually milder and not disabling. They also noted that this group of neurodevelopmental disorders shares a high level of comorbidity; for example, up to 38% of patients with a diagnosis of autism also fulfill criteria for other developmental disorders, such as bipolar disorder, ADHD and OCD [6, 7].

Seven years ago, when many believed that postnatal factors, such as vaccines, were probable causes of autism, a number of researchers hypothesized that prenatal origins were more likely. Data from existing twin studies and an increased incidence within families showed that autism had genetic components. However, the prevalence of this disorder was increasing at an accelerated rate, faster than could be explained by genetic mutations alone. Furthermore, postmortem brain studies showed abnormalities in structures that develop before birth, and extensive epidemiologic research did not support causation by vaccines. It is now apparent that autism – as well as other related neurodevelopmental disorders – may involve multiple causal factors, and that their origins are, in most cases, prenatal.

Until several years ago, only a small number of studies suggested that prenatal interactions between genes and the environment might cause autism, schizophrenia, and other neurodevelopmental disorders. There were relatively few researchers taking this approach, and they were working in disciplines that were historically disconnected from one another. This was a new concept, and important questions arose from this idea and included *when, how and for how long* interactions between genes and the environment could occur, to result in a neurodevelopmental disorder such as autism. It became clear through the existing scientific literature and clinical observations that the *fetal environment* should be explored as the staging ground for neurodevelopmental disorders.

Historically, research in neurodevelopmental disorders has focused on single genes and biomarkers. In 2003, a small group of investigators envisioned shared elements and complex prenatal origins in causation of these disorders and perceived that environmental influences at multiple levels can act on genetic vulnerabilities to disrupt normal brain development. They predicted that epigenetics and normal gene variants (polymorphisms) would be important contributors to neurodevelopmental disorders. Further, they understood that numerous abnormalities of brain development that occur in neurodevelopmental disorders are not abnormalities of *form*, but rather, disorders of *function* in neurons, neuroglia and their circuitry that affect *future* responses and performance in these tissues. Therefore, dysfunctions that result will often not affect parameters measured at birth, such as Apgar scores or the standard neonatal exam, but will later reveal themselves in symptoms of neurodevelopmental disorders during early childhood, adolescence, or young adulthood. They also believed that changes that occur during fetal life could not only affect the brain, but could also cause system-wide changes in cellular physiology during postnatal development.

The Fetal Physiology Foundation was started to support the concept that the prenatal origins and biological complexity of neurodevolopmental disorders are results of environmental factors acting on genetic susceptibility. This broad new approach was proposed in order to create dialogue among researchers across disciplines, and was based on exploration of development at the cellular level during fetal life. Through the vehicle of this nonprofit research organization, investigators could find both the forum and financial support they needed for small, novel projects centered on fetal neurodevelopment that would lead to larger basic and translational studies.

In 2006, the Fetal Physiology Foundation held its inaugural symposium entitled *Fetal Mechanisms in Neurodevelopmental Disorders* at The Kennedy Krieger Institute in Baltimore, Maryland. Participants identified and discussed both recognized and hypothetical prenatal cellular mechanisms responsible for abnormal neurodevelopmental trajectories. Topics in the symposium illustrated the innovative research the Fetal Physiology Foundation plans to facilitate and support [8].

Since then, research has continued to suggest that the fetal environment is the staging ground for neurodevelopmental disorders and that these disorders result, in part, from genetic susceptibility influenced by various factors during prenatal life. It is likely that the overlapping symptoms among these disorders result from shared fetal mechanisms that interfere with normal cell programming at critical periods during gestation, and that this process is due to multiple environmental, genetic, maternal, and even transgenerational factors. A number of these influences can affect the *intrauterine environment*, such as maternal stress, endocrine alterations, immune responses to infection or a foreign (e.g., paternally determined) protein expressed in the fetus, as well as exposure to pesticides and medications.

Because gene-environment interplay has its most significant effects on brain development within the intrauterine environment, the Fetal Physiology Foundation believed that maternal influences on the fetal environment would be an important topic of investigation. This genre of research, though expanding, was a relatively new approach to the causes of neurodevelopmental disorders. In 2008, the Fetal Physiology Foundation held its second symposium: *Maternal Influences on Fetal Neurodevelopment*, at The Johns Hopkins University School of Medicine, supported by Kennedy Krieger Institute and the *Eunice Kennedy Shriver* National Institute of Child Health and Human Development and moderated by Tonse Raju, M.D. Participants included both practicing physicians and researchers from diverse disciplines whose work involves maternal–fetal interaction. The chapters of this book, written by invited speakers at the symposium, represent the relevant research presented and discussed. They characterize ongoing efforts of The Fetal Physiology Foundation to foster understanding and support research into maternal and fetal mechanisms that lead to neurodevelopmental disorders.

Baltimore, MD — Andrew W. Zimmerman
Boston, MA — Susan L. Connors
Baltimore, MD — Rosa M. Dailey

References

1. Autism and Developmental Disabilities Monitoring Network Surveillance Year 2006 Principal Investigators, Centers for Disease Control and Prevention (CDC) (2009) Prevalence of autism spectrum disorders – Autism and Developmental Disabilities Monitoring Network, United States, 2006. MMWR Surveill Summ 58(10):1–20
2. Moreno C, Laje G, Blanco C, Jiang H, Schmidt A, Olfson M (2007) National trends in the outpatient diagnosis and treatment of bipolar disorder in youth. Arch Gen Psychiatry 64(9):1032–1039
3. Pastor PN, Reuben CA (2002) Attention deficit disorder and learning disability: United States 1997–1998. Vital Health Stat 10(206):1–12
4. Merikangas KR, He JP, Brody D, Fisher PW, Bourdon K, Koretz DS (2010) Prevalence and treatment of mental disorders among US children in the 2001–2004 NHANES. Pediatrics 125(1):75–81
5. Pessah IN, Lein PJ (2008) Evidence for environmental susceptibility in autism. In: Zimmerman AW (ed) Autism: current theories and evidence. Humana Press, Totowa, NJ, pp 409–428
6. Goldstein S, Schwebach AJ (2004) The comorbidity of pervasive developmental disorder and attention deficit hyperactivity disorder: results of a retrospective chart review. J Autism Dev Disord 34(3):329–339
7. Stahlberg O, Soderstrom H, Rastam M, Gillberg C (2004) Bipolar disorder, schizophrenia, and other psychotic disorders in adults with childhood onset AD/HD and/or autism spectrum disorders. J Neural Transm 11(7):891–902
8. Connors SL, Levitt P, Matthews SG, Slotkin TA, Johnston MV, Kinney HC, Johnson WG, Dailey RM, Zimmerman AW (2008) Fetal mechanisms in neurodevelopmental disorders. Pediatr Neurol 38(3):163–176

Contents

Contributors

Steven Buyske, Ph.D.
Department of Statistics & Biostatistics, Rutgers University, Piscataway, NJ, USA

Susan L. Connors, M.D.
Departments of Medicine and Pediatrics, Massachusetts General Hospital, Boston, MA, USA; Department of Neurology and Developmental Medicine, Kennedy Krieger Institute, Baltimore, MD, USA

Rosa M. Dailey, B.A.
President and Founder, Fetal Physiology Foundation, Baltimore, MD, USA

Janet A. DiPietro, Ph.D.
Department of Population, Family and Reproductive Health, Johns Hopkins University, 615 N. Wolfe St., E4531, Baltimore, MD 21205, USA

Christopher S. Ennen, M.D.
Division of Maternal-Fetal Medicine, Department of Gynecology & Obstetrics, Johns Hopkins University School of Medicine, Baltimore, MD, USA

Laura M. Glynn, Ph.D.
Department of Psychology, Chapman University, Department of Psychiatry and Human Behavior, University of California, Irvine, 333 The City Blvd. W, Suite 1200 Orange, CA 92668, USA

Ernest M. Graham, M.D.
Maternal-Fetal Medicine Division, Johns Hopkins Hospital, Phipps 228, 600 N. Wolfe St., Baltimore, MD 21287, USA

Elaine Hsiao, B.S.
Biology Division, California Institute of Technology, Pasadena, CA, USA

William G. Johnson, M.D.
Department of Neurology, UMDNJ-Robert Wood Johnson Medical School, Hoes Lane, Piscataway, NJ, 08854, USA and
Center for Childhood Neurotoxicology & Exposure Assessment, UMDNJ-Robert Wood Johnson Medical School, 671 Hoes Lane, Piscataway, 08854, NJ, USA

George H. Lambert, M.D.
Division of Pediatric Pharmacology and Toxicology, Department of Pediatrics, UMDNJ-Robert Wood Johnson Medical School, Piscataway, NJ, USA

Paul H. Patterson, Ph.D.
Biology Division, California Institute of Technology, Pasadena, CA, USA

Tonse N.K. Raju, M.D., DCH
Eunice Kennedy Shriver National Institute of Child Health and Human Development, 6100 Executive Blvd, Rm 4B03, Bethesda, MD 20892, USA

Joanne F. Rovet, Ph.D.
Hospital for Sick Children, University of Toronto, 555 University Avenue, Toronto, ON, M5G1X8, Canada

Stephen E.P. Smith, Ph.D.
Biology Division, California Institute of Technology, Pasadena, CA 91125, USA; Departments of Neurology and Pathology, Harvard Medical School, Beth Israel Deaconess Medical Center, 330 Brookline Avenue, E/CLS-717, Boston, MA 02215, USA

Edward S. Stenroos, B.S.
Department of Neurology, UMDNJ-Robert Wood Johnson Medical School, Piscataway, NJ, USA

E. Fuller Torrey, M.D.
Medical Research Institute, Chevy Chase, Bethesda, MD, USA

Karen A. Willoughby, M.A.
Department of Psychology, University of Toronto, Toronto, ON, Canada

Robert Yolken, M.D.
Stanley Division of Developmental Neurovirology, Johns Hopkins School of Medicine, 600 N. Wolfe Street, Blalock 1105, Baltimore, MD 21287-4933, USA

Andrew W. Zimmerman, M.D.
Department of Neurology and Developmental Medicine, Kennedy Krieger Institute, Baltimore, MD, USA; Departments of Neurology, Psychiatry and Pediatrics, Johns Hopkins University School of Medicine, Baltimore, MD, USA

Chapter 1
Brave New World: The Intrauterine Environment as the Biological Foundation for the Lifespan

Tonse N.K. Raju

Keywords Developmental origins of adult diseases • Fetal behavior • Fetal programming • Maternal hypothyroxinemia • Maternal–fetal interface • Perinatal encephalopathy

It is widely recognized that Sir Joseph Barcroft (1872–1947) laid the modern methodological foundation for the study of the mammalian fetus. After studying high-altitude physiology for decades, Barcroft, at age 60, turned his attention to the physiology of the mammalian fetus to learn how it develops in an environment of extremely low oxygen tension, or as he put it, "while living on Mt. Everest In-utero." Although his focus was on fetal physiology, he never lost sight of the fact that "…one day, the call will come and the fetus will be born. Not only has the fetus to develop a fundamental life… [to withstand] the shock of birth, but to [also survive in] its new environment [1]." Thus, he implied that physiological processes in the fetus need to be interpreted with a perspective for long-term survival.

More than six decades since the publication of his book, "Research on Prenatal Life" [1], the study of the mammalian fetus and its environment has grown into a robust, multidisciplinary science that confirms Barcroft's visionary statement. The intrauterine environment not only prepares the fetus to withstand the "shock of birth," but also shapes the life course of the infant and his or her mother. In fact, bidirectional developmental programming prepares the maternal–fetal dyad for its interactions and life journey together.

Knowledge in this field has grown rapidly, largely due to unprecedented advances in technology and research methods, and collaboration among scientists from diverse disciplines previously considered "unrelated" fields. Adaptation of research methods from physiology, developmental neuroscience, child psychology,

T.N.K. Raju (✉)
Eunice Kennedy Shriver National Institute of Child Health and Human Development, 6100 Executive Blvd, Rm 4B03, Bethesda, MD 20892, USA
e-mail: rajut@mail.nih.gov

A.W. Zimmerman and S.L. Connors (eds.), *Maternal Influences on Fetal Neurodevelopment: Clinical and Research Aspects*,
DOI 10.1007/978-1-60327-921-5_1,

molecular biology, population genetics, biomedical imaging, evolutionary biology, and epidemiology was essential to gain insight into the complex world of the intrauterine environment.

The chapters in this volume present a brief overview of the state of the science on this topic. Some major themes addressed by the authors are summarized below.

The Developing Brain

Medical and biology students for generations were taught that the newborn infant's brain is essentially a blank slate onto which, gradually, new knowledge is "filled-in [2]." Some students and teachers may hold such views even now.

The collective evidence, however, seriously contradicts the notion that the mammalian newborn is no more than a "brainstem creature." Within seconds after birth, the newborn infant unleashes a large set of sophisticated neural processing pathways to collect and evaluate, collate and assemble, prune and consolidate neurosensory inputs, constantly remodeling his or her own brain [2]. The infant continues to observe, explore, imagine, and learn more than we ever thought possible. It has been said that a newborn infant is a "scientist in the crib [2]."

In fact, preparations for sculpting the infant brain start long before birth. In addition to the continuous supply of nourishment to the fetus from the mother, there is constant interaction between the two throughout pregnancy. The routes for such interactions include the placenta, the fetal membranes, the amniotic fluid and the uterine wall. The mediators for such intense interplay might be a rich array of hormones, biochemical substances, and cellular intermediates. There may be other channels and mediators yet to be explored. In a large sense, these efforts are geared to assure that the infant launches successfully into this noisy world of light, ambient air, and atmospheric pressure, and to prepare its mother for the role of continuing caregiver.

In Chapter 3, DiPietro summarizes the evidence from several sources on how maternal psychological functioning modulates fetal neurobehavior and evokes responses from the autonomic nervous system. Maternal mood and anxiety states have been shown to elicit responses in fetal motor activity, changes in heart rate and its variability, and breathing. It is remarkable that fetal responses, in turn, influence the mother's biology, often without reaching the level of her conscious perception. In DiPietro's studies, maternal heart rate and skin conductance responses to fetal motor activities at different stages of pregnancy were remarkably similar, whether the mother was from Baltimore, or from Lima, Peru. This apparent universality implies that the phenomenon is biological, rather than cultural. The reasons behind the evolutionary need for such a response are yet to be understood.

Glynn describes an integrated approach to appreciating the complex interrelationships within the maternal–fetal dyad. By means of maternal and fetal programming, both the fetus and the maternal prenatal milieu constantly influence each other, and affect the development of the fetal brain. In addition, we learn of animal studies in which permanent alterations in maternal brain structure and function take place during pregnancy. Such changes facilitate and strengthen maternal–offspring bonding, perhaps through hormonal mediators.

Disorders in the Offspring and the Intrauterine Environment

In Chapter 5, Rovet and Willoughby review the role of maternal thyroid function in promoting fetal brain development. During the first half of gestation, the human fetus depends entirely upon the mother for thyroid hormones, as shown by measurable concentrations of T4 in the fetal brain long before the fetal thyroid gland begins secreting these hormones. Deprived of the neurodevelopmental effects of maternal thyroxine during gestation, children treated for congenital hypothyroidism secondary to untreated maternal hypothyroidism do not develop full intellectual function, even when therapy is started soon after birth [3].

However, controversy exists concerning the value of universal thyroid screening during pregnancy and the antenatal treatment of subclinical hypothyroid states. Some reasons for this uncertainty are as follows. The serum concentrations of thyroid hormones fluctuate in a somewhat unpredictable manner during pregnancy. Thus developing a standard definition for "subclinical" hypothyroidism is difficult. The laboratory methods and standards for measuring thyroid hormones vary greatly, thus a given blood sample tested in two laboratories may report different results. Perhaps due to the above reasons, no clinical trial to date has shown benefits from intervention for subclinical maternal hypothyroidism during pregnancy.

The US Clinical Trials registry lists many ongoing and recently completed clinical trials on this topic. In one such trial [4] (scheduled to end in 2015), the Maternal–Fetal Medicine Network of the *Eunice Kennedy Shriver* National Institute of Health and Human Development (NICHD), is testing the effect of thyroxine therapy for subclinical hypothyroidism or hypothyroxinemia diagnosed during the first half of pregnancy. In a double-masked[1] randomized controlled trial, 1,000 women have been enrolled and the IQ of the child at 5 years of age is the primary outcome. Recruitment is complete and infant follow-up is continuing.

Ennen and Graham review the complex topic of perinatal asphyxia and neonatal brain damage. Recent demonstration of the beneficial effects of mild therapeutic hypothermia for infants with severe perinatal encephalopathy [5], and the reduced incidence of cerebral palsy in preterm infants with antenatal exposure to magnesium sulfate are two encouraging advances in this field. Magnesium sulfate is the first pharmacological agent with a potential for preventing cerebral palsy.

In Chapters 7 and 8 Smith et al. and Yolken et al. address some of the most perplexing questions in neurobiology, namely, what is the role of intrauterine infections on later development of schizophrenia, and does an altered or "activated"

[1]Previously referred to as "double-blind."

maternal immune system lead to autism spectrum disorders during childhood? The microbial culprits implicated in the etiology of schizophrenia include *Toxoplasma Gondii*, herpes, rubella, polio, measles, and influenza viruses. Activation of the maternal immune system by prenatal maternal infections, and the adverse effects of inflammatory mediators may be causally related to poor neuropsychiatric outcomes in offspring in this situation.

Johnson and colleagues describe the role of maternally acting gene alleles in causing neurodevelopmental disorders – a novel and rapidly expanding field of science. At least 35 distinct neuropathological conditions have been identified in which the mother is the "patient," and her offspring develop abnormally as a result of her maternally acting alleles. The conditions include autism, Down syndrome, rheumatoid arthritis, schizophrenia, and spina bifida among others. This field of genetics is new in neurodevelopmental disorders. In addition to the mother, maternal grandparents may contribute "teratogenic alleles" responsible for such conditions as schizophrenia and autoimmune disorders that develop years later in the children.

Future Research

The state-of-the-science reviews in this volume also identify gaps in our knowledge and suggest further research to fill them. A few additional proposals are discussed below.

What, How and Why? Most studies in this field have been exploratory. Studies designed to "*see what happens when* ..." help generate hypotheses to be tested. "How" and "why" may be the obvious next type of questions that need to be asked. At present, genetic and molecular models have been used to seek mechanistic explanations for the observed responses to changes in the intrauterine environment. However, to address the more difficult "why" questions, one needs to design long-term prospective studies. Imaginative techniques may need to be developed by integrating research methods from biomedical and bioengineering sciences, as well as from anthropology and evolutionary biology.

Developmental, Not Only Intrauterine, Environment

Development is nonlinear and each system follows its own trajectory while interacting with other systems. The intrauterine period is one of many phases in the lifespan from an embryo and fetus to adulthood and old age. There are critical phases of growth and maturation, and of vulnerability and plasticity. Thus, a broader approach to understanding "developmental" environmental influences on adult phenotypes may be the next frontier in this field of research.

Public Health Implications of Developmental Programming

The emerging science of developmental origins of adult diseases (DOAD) has fundamentally altered our approach to adult onset disorders, such as type II diabetes, coronary artery disease (CAD), stroke, hypertension, and obesity. What are the public health implications of the evolving knowledge of DOAD? Can we develop interventions based on what we have learned? For instance, low birth weight has been associated with adult onset CAD. To reduce the risk of CAD, therefore, should we implement interventions to optimize fetal growth? Is there a risk of increasing childhood obesity by such an approach? Similarly, if maternal infections are proximate etiological "causes" for neuropsychiatric conditions, should one develop methods to diagnose them early and treat the fetus? These are but two examples in this rapidly growing field.

Manipulating the Environment to "Optimize" Outcomes

This field opens up a new world of possibilities and questions. As an example, can one manipulate the intrauterine environment to improve or modify maternal–infant bonding, or prevent damage from adverse environmental pollutants? Are there ethical limits to such approaches?

Intrauterine Environmental Deprivation

Preterm infants will be deprived of the full complement of intrauterine environmental influences compared to their term-born counterparts. The negative consequences from such deprivations, if any, could impact a large segment of the population, since the preterm birth rate has been increasing in the USA, reaching an all time high of 12.8% in 2006 [6].

Depending upon the extent of prematurity, preterm infants are at two- to tenfold higher risk than term infants for cerebral palsy, sensory motor impairments, seizures, learning and behavioral problems, and cognitive and psychological dysfunctions, many of which persist into adulthood [7]. The proportion of such morbidities attributable to the early termination of the influences of the intrauterine environment are unclear.

Kinney et al. showed that brain weight increases by about 35% between 35 and 40 weeks of gestation [8]. What are the intrauterine forces that trigger the rapid rate of synaptogenesis and dendritic arborization necessary for the late gestational surge in brain growth? Might an early termination of the intrauterine environment be responsible for disrupting those processes and the brain's growth spurt, leading to a higher proportion of developmental and learning disabilities, and psychiatric dysfunctions in late preterm infants born between 34 and 36 weeks of gestation? Answers to these and related questions remain to be addressed in future studies.

Fetal Learning

The legend of Abhimanyu in the Indian epic *Mahabharata* may be the first reference to fetal learning (or programming?). While he was still in his mother's womb, Abhimanyu learned from his father Arjuna, the secret art of penetrating the deadly *Chakravyuha*, or circular formation of infantrymen, archers, horse-drawn chariots, and elephants in battles. Arjuna was describing this secret to his pregnant wife, and midway through the narration she dozed off. Thus the fetus could not learn how to get out of the Chakravyuha. Years later, this half-knowledge would cost young Abhimanyu his life. In the Great Battle, Abhimanyu penetrates the circular formation, but, not knowing how to get out of it, he is trapped and killed by the enemy.

Although a legend, Abhimanyu's story raises interesting questions: can maternal learning during pregnancy help program the fetus to learn, too? What is the nature of such knowledge?

There may be a risk, however, in taking the notion of fetal learning too far. As a recent article in *The Washington Post* describes, dozens of products have flooded the market as "prenatal learning systems," enticing pregnant women to enroll their unborn children in womb-schooling [9]. None of these products has been tested for efficacy, and none has received the approval of the Food and Drug Administration as a medical device. As well as being ineffective, there is potential that they may cause harm. The processes for in-utero learning (or conditioning) might have evolved through natural selection for reasons yet to be understood. Thus, it is premature for us to become "fetal teachers," and to develop devices for enhancing fetal learning systems. It may be wise to follow the dictum "not to fool Mother Nature."

Conclusions: View from Mt. Barcroft

Our current knowledge about the intrauterine environment could not have materialized without the groundbreaking work of Barcroft and his contemporaries in the early- to mid-twentieth century [1], along with technological advancements during the past 20 years. Barcroft began writing his book in 1939 as Great Britain entered World War II [1], and had plans to write Part II devoted to the developing nervous system. Unfortunately, a few weeks after receiving the printed copy of Part 1, Barcroft died of a heart attack on March 21, 1947.

In 1954, the US Board on Geographic Names christened a 13,040-ft peak on California's White Mountains, *Mt. Barcroft* [10]. Several laboratory facilities have been built by the University of California on these mountain peaks to conduct

Fig. 1.1 Nello Pace Laboratories on the Eastern slope of Mt. Barcroft (*to the left*)

research on diverse topics, including mammalian and plant physiology, ecology, geology, astrophysics, chemistry, and global climate [11]. On the eastern slope of Mt Barcroft is the Nello Pace Laboratory with facilities for conducting fetal physiology research literally on the mountain (Fig. 1.1). How far can one see into the future of research in this field from the top of Mt. Barcroft? The answer may be limited only by one's imagination.

Acknowledgment I wish to thank Daniel Pritchett and Frank L. Powell, Ph.D, White Mountain Research Station, for providing me with the Figure.

References

1. Barcroft Sir J (1947) Research on prenatal life. Part 1. Blackwell Scientific, Oxford
2. Gopnik A, Meltzoff AN, Kuhl PK (1999) The scientist in the crib: minds, brains, and how children learn. William Morrow, New York
3. Rose SR, Brown R, The American Academy of Pediatrics, section on endocrinology and committee on genetics (2006) Update of newborn screening and therapy for congenital hypothyroidism. Pediatrics 117:2290–2303
4. Thyroid therapy for mild thyroid deficiency in pregnancy (TSH) (2009) http://clinicaltrials.gov/ct2/show/NCT00388297?term=thyroid+AND+pregnancy&rank=2. Accessed 28 Sep 2009
5. Shankaran S (2009) Neonatal encephalopathy: treatment with hypothermia. J Neurotrauma 26:437–443
6. Martin JA, Hamilton BE, Sutton PD, Ventura SJ, et al. Births: Final data for 2006. National vital statistics reports; vol 57 no 7. Hyattsville, MD: National Center for Health Statistics. pp 2–102, 2009

7. Moster D, Lie RT, Markestad T (2008) Long-term medical and social consequences of preterm birth. N Engl J Med 359:262–273
8. Kinney HC (2006) The near-term (late preterm) human brain and risk for periventricular leukomalacia: a review. Semin Perinatol 30:81–88
9. Salslow R (2009) Pre-preschool: new devices aim to help babies start learning before birth. But are they just a lot of noise? The Washington Post, pp E1–E5
10. Mt. Barcroft (2009) http://geonames.usgs.gov/. Accessed 28 Sep 2009
11. White Mountain Research Station. http://www.wmrs.edu

Chapter 2
In the Beginning

Janet A. DiPietro

Keywords Fetal programming • Prenatal development • Fetal neurobehavior • Fetal heart rate • Fetal movement

The explosive rate of growth and development that occurs during the period before birth is unparalleled at any other point in the lifespan. In just 266 days, a single fertilized cell develops into a sentient human newborn infant. While information regarding the structure of the developing embryo and fetus has long been available, knowledge concerning prenatal development of function is more recent. The advent of real-time ultrasound and improvements in electronic fetal heart rate monitoring technology in the early 1980s were followed by a wave of research on fetal neurobehavioral development. A renewed surge of interest in the prenatal period as the foundation for later life has recently been fostered by enormous attention devoted to "fetal programming" in relation to later health and well-being. The concept of fetal programming has been applied broadly to represent discoveries of prenatal influences on postnatal conditions, typically with adult onset [1–4]. This avenue of research considers the role of maternal and fetal factors on subsequent organ function, including the brain and nervous system, using an epidemiologic framework to study related morbidity and mortality.

The assumption that earlier circumstances, including those during the prenatal period, affect later development has been at the core of developmental science since its inception. Thus, the basic foundations of developmental sciences that are concerned with formation and expression of individual differences, along with the moderating role of early environmental influences, have begun to converge with epidemiologic methodology. Yet to be reconciled, however, is the disparity between the nature of the data and the underlying assumptions. While fetal programming research has generated an enormous body of data, most studies rely on readily

J.A. DiPietro (✉)
Department of Population, Family and Reproductive Health, Johns Hopkins University, 615 N. Wolfe St., E4531, Baltimore, MD 21205, USA
e-mail: jdipietr@jhsph.edu

A.W. Zimmerman and S.L. Connors (eds.), *Maternal Influences on Fetal Neurodevelopment: Clinical and Research Aspects*,
DOI 10.1007/978-1-60327-921-5_2, © Springer Science+Business Media, LLC 2010

available variables, such as birth weight, that provide only a vague approximation of the gestational environment and can offer little information about mechanisms that mediate the observed associations. In contrast, the developmental approach, traditionally implemented within fields such as psychology, psychobiology, and neuroscience, relies on discovery to generate useful indicators of nervous system development and then applies them to considerably smaller study samples. The emphasis is on *measurement of function* in an effort to approximate the underlying physiological substrate as closely as possible. By focusing on a single facet (i.e., neural function) that encapsulates development from the prenatal to postnatal period, research is directed at evaluating how early experiences or exposures might affect the latter via their influence on the former.

Fetal Neurobehavioral Development

Developmental parameters that are measured extensively in the neonate and infant, and are integral to theories of development, originate neither at term nor with birth [5, 6]. Development during the prenatal period proceeds along a continuum, with behaviors becoming incrementally more complex and varied as gestation proceeds. Like all other developmental stages, the fetal period is not monophasic or uniform; behavior in the early fetal period is largely reflexive and involves the entire body, while behavior near term is far more fluid, integrated, and distinct. Fetal neurobehavioral research typically centers around four measures which had previously been established as core neonatal and infant parameters. These include fetal heart rate and variability, fetal motor behavior and activity level, fetal state development and profiles (essentially the interaction of heart rate and motor activity plus fetal eye movements), and fetal detection of and responsiveness to stimulation provided by the environment. Initial research was devoted to developing normative data about fetal neurodevelopment in these domains [7–12]. More recent longitudinal studies continue to document normal ontogeny [13–15].

In the mid-1990s, the National Institute of Child Health and Human Development convened a series of conferences to integrate knowledge generated by obstetric and developmental research, with the goal of advancing methods to measure neurobehavior in the fetus as indicators of nervous system development [16]. The theory that the neurobehavior of the fetus provides information regarding neurological development has been supported by evidence from studies conducted in healthy populations [17–21]. Further support is provided by observations of differences in neurobehavioral functioning in at-risk fetuses. Studies have indicated that fetuses afflicted by congenital anomalies related to the nervous system [22–25], or who express growth restriction [26], show different developmental trajectories. Prenatal neurobehavioral alterations have also been documented in fetuses exposed to other deleterious maternal conditions, including maternal diabetes [27], substance use [28, 29], and maternal hypothyroidism (see Chapter 5, "Maternal Thyroid Function During Pregnancy: Effects on the Developing Fetal Brain").

Models of fetal neurodevelopment typically accept that the refinement of fetal parameters during gestation parallels and reflects the enormous growth of the developing nervous system. Understanding of fetal developmental trajectories has value both in ascertaining normal ontogeny and in evaluating how deviations from typical development can affect postnatal outcomes. Figure 2.1 provides a schematic illustration of how fairly minor deviations from the norm early in development can become amplified over time and continue through the postnatal period. Development during the embryonic and earlier fetal periods is more canalized or restricted than later development. Although it is likely that the roots of individuality are present earlier than midgestation, significant variation in their neurobehavioral expression becomes more pronounced as gestation advances. Deviations from normal may be the result of either constitutionally determined, inherent characteristics of the individual fetus (such as those that are genetic) or the result of exposures to maternal or environmental factors or both. In general, alterations to typical developmental trajectories that begin closer to the origin of any developmental process have the most significant repercussions; thus, the fetal period provides prime opportunity for potential interventions that may have long-lasting and positive impacts. The remainder of this chapter is focused on studies that evaluate whether elements of fetal neurobehavior predict postnatal development, either within or across functional domains.

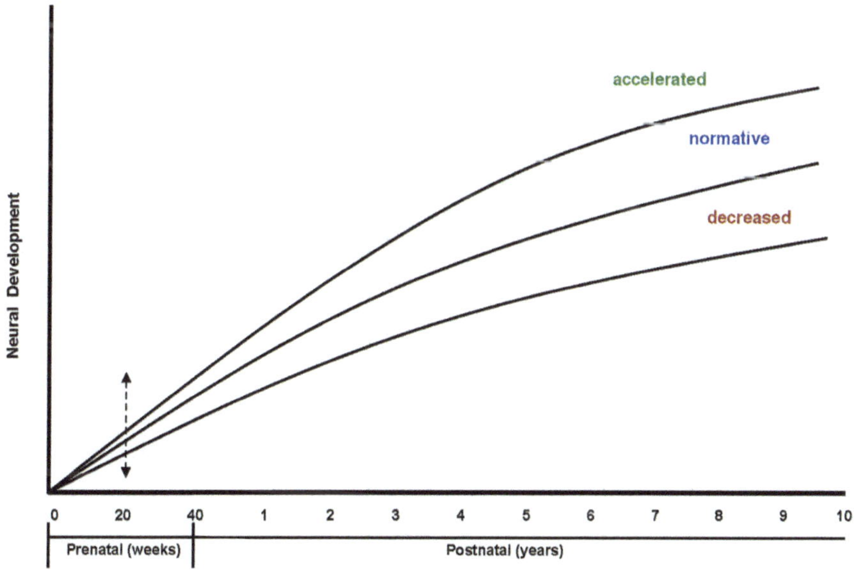

Fig. 2.1 Schematic illustration of accelerated, normative, and decreased developmental trajectories during the fetal period and into childhood. Note that the magnitude of developmental disparity becomes amplified with age

Prenatal to Postnatal Continuities

The initial step in evaluating prenatal to postnatal continuity of neurobehavior is to establish that measurable fetal parameters vary among fetuses and show stability during gestation within individuals. That is, individuals who express high (or low) levels of a fetal neurobehavioral characteristic at one point during gestation should express high (or low) levels of that characteristic when measured again. Such stability has been shown during the fetal period. Measures of fetal neurodevelopment stabilize during gestation in a hierarchical fashion, commencing with the most basic, fetal heart rate at baseline conditions, and progressing to the ability to mount more complex coordinated responses to stimulation [30]. Most, but not all, studies tend to commence at midgestation (approximately 20 weeks) since it is difficult to collect continuous fetal heart rate data prior to that time. Thus, it is possible that stability in fetal neurodevelopment commences earlier than 20 weeks of gestation.

Predictability from the prenatal to postnatal period can reflect both conservation of a similar attribute (e.g., prenatal activity level to postnatal activity level) and cross-domain associations (e.g., prenatal activity level to postnatal difficult temperament), which may share another underlying attribute (e.g., regulatory control) but are expressed by different behavioral manifestations. Regardless of the nature of the research question, methods that are required for viewing and monitoring the fetus, a nonbreathing organism surrounded by fluid and positioned with head down in a tightly circumscribed space during much of gestation, are quite different from those that can be applied for collecting data on individual infants and children who can be both directly seen and examined. The technical difficulties in conducting this type of research are daunting, and consequently, the existing literature is not large. Nonetheless, studies have consistently been able to detect prenatal to postnatal continuities, and while the magnitude of the correlations tends to be modest, they are consistent in size and yield predictive relationships to infancy and early childhood.

Within Domain Relations

Fetal heart rate and heart rate variability are the most stable individual characteristics during gestation and remain correlated with infant heart rate parameters through at least the first year of life [31, 32]. There is a statistically significant relationship between slow prenatal and postnatal heart rate at age 10 [33]. With respect to motor activity, greater fetal motor activity is associated with greater infant motor activity in the neonatal period [34], at 3 and 6 months postpartum [30] and at age 2 [35], although the latter finding was true only for boys. More recently, a study of twins has revealed that the more active twin in utero, assessed prior to 14 weeks gestation, remained the more active twin at 6 months postpartum showing significant prenatal to postnatal stability in this measure [36]. Several studies have also shown that stable elements of behavioral states continue from the prenatal to postnatal periods [37, 38]. For example, fewer night wakings occur in 3-month-old infants who had higher than average levels of fetal state organization [30].

Cross-Domain Associations

Much attention has been devoted to evaluating prenatal to postnatal continuities that span different domains in predictable ways. A link between greater fetal motor activity and motor development at 6 months of age was among the findings reported from the Fels Longitudinal Study in the 1930s [39]. Recently, this has been partially confirmed during the neonatal period in that newborns who displayed higher levels of motor activity as fetuses showed more optimal motor development and reflexes [40], suggesting that either fetuses that move more have better motor control or that fetal motor activity provides a practice effect that develops musculature.

Fetal reactivity to stimuli has been evaluated as a predictor of both cognitive and temperamental outcomes. Prenatal reactivity is assessed in two distinct ways. The first is by applying a vibroacoustic stimulus to or near the maternal abdomen and evaluating fetal responsiveness or habituation to it. The second is to stimulate a maternal emotional response which, in turn, can elicit a fetal response via alterations to the intrauterine environment. Fetal habituation proficiency has been identified as an indicator of subsequent advanced mental development at age one [10] and also as a predictor of visual recognition memory, a measure of information processing, at 6 months [41]. In contrast, emotional regulation in infants has been linked to fetal responsiveness to induced maternal physiological arousal. Greater fetal heart rate responsiveness to a cognitive challenge presented to the pregnant woman was predictive of greater motor reactivity to a standard novelty paradigm and a trend for greater infant negativity [42]. Similarly, fetuses that displayed more fetal heart rate reactivity (as well as motor reactivity) to maternal viewing of a labor and delivery film were more irritable to the manipulations encountered in a neurodevelopmental exam at 6 weeks [43].

Evaluation of heart rate patterning as an indicator of neural development within the autonomic nervous system in infants and children has a distinguished history in developmental research. Spontaneous fetal heart rate is the easiest to collect and most reliable fetal neurobehavioral measure. Predictive correlations between baseline, undisturbed fetal heart rate and variability, quantified by a variety of methods, have been established with various aspects of postnatal function. Significant associations have been reported between higher fetal heart rate and lower threshold to novelty [44], emotional tone [30], and, paradoxically, positive reactivity [42] in early infancy.

Two studies report that indicators of higher fetal heart rate variability as well as steeper developmental trajectories for this measure during the second half of gestation are positively associated with mental, psychomotor, and language development [45] and symbolic play proficiency [46] during the third year of life. Psychomotor development at 18 months has also been predicted by components of fetal heart rate variability [47]. Most recently, fetal heart rate variability and its trajectory have been positively associated with brainstem auditory evoked potential (BAEP) activity in neonates, a measure of the speed of neural conduction within portions of the auditory pathway [40]. In addition, fetal somatic–cardiac coupling, an indicator of the integration between neural pathways controlling heart rate and motor activity, has also been associated with these evoked potentials. These results suggest that autonomic maturation in utero may reflect elements of centrally mediated maturation of the nervous system in general.

Summary

An accumulating body of evidence points to functional continuity between the prenatal and postnatal periods and reveals the spectrum of normal variation in underlying neural integrity among individuals. Perhaps no other developmental period, beyond the fetal one, yields the same potential to reveal the complexities of human ontogeny, yet no other period of developmental inquiry is so heavily dependent on technology to answer even the most basic of questions. As a result, the current literature evaluating fetal-to-child continuities is fairly small and not uniform in its findings or methods, and there have been few attempts at replication. However, given the vast differences in the circumstances of the fetus compared to the child, it can be argued that the odds strongly favor Type II statistical errors in studies, i.e., finding no significant relationships when one actually exists. As such, detection of any significant associations might be considered noteworthy. As observational studies continue to accrue, replicated and extended findings will coalesce and serve to inform us, while single instance findings that cannot be replicated will be discarded as spurious. New technologies, such as 4-D ultrasound, will change the nature of research questions we are able to address. In addition, maternal health, the uterine environment, and genetic susceptibility associated with neurodevelopment have become areas of intense research interest. Immune activation during maternal illness (see Chapter 7, "Activation of the Maternal Immune System as a Risk Factor for Neuropsychiatric Disorders", Smith) or occult infection (see Chapter 8, "Prenatal Infections and Schizophrenia in Later Life-Focus on *Toxoplasma gondii*") and subclinical endocrine disorders such as maternal hypothyroidism (see Chapter 5, "Maternal Thyroid Function During Pregnancy: Effects on the Developing Fetal Brain") provide further contributions to alterations of neurodevelopmental trajectories. Maternal genes that predispose to neurodevelopmental disorders in the fetus most likely do so by altering the uterine environment with regard to the availability of nutrients and the intensity of maternal immune reactions, among other mechanisms (see Chapter 9, "Maternally Acting Alleles in Autism and Other Neurodevelopmental Disorders: The Role of HLA-DR4 Within the Major Histocompatibility Complex (MHC)"). While current empirical support leaves little doubt as to the importance of the prenatal period in providing the foundation for postnatal life, the opportunity for discovery – and the potential benefits of newly acquired knowledge – remains boundless.

References

1. Barker DJP (2006) Adult consequences of fetal growth restriction. Clin Obstet Gynecol 49:270–283
2. Phillips D (2001) Programming of adrenocortical function and the fetal origins of adult disease. J Endocrinol Invest 24:742–746
3. Szitanyi P, Janda J, Poledne R (2003) Intrauterine undernutrition and programming as a new risk of cardiovascular disease later in life. Physiol Res 52:389–395
4. Young J (2002) Programming of sympathoadrenal function. Trends Endocrinol Metab 13:381–385
5. Als H (1982) Toward a synactive theory of development: promise for the assessment and support of infant individuality. Infant Ment Health J 3:229–243
6. Prechtl HFR (1984) Continuity and change in early neural development. In: Prechtl H (ed) Continuity in neural functions from prenatal to postnatal life. J.B. Lippincott, Philadelphia, PA, pp 1–15
7. Dalton K, Dawes GS, Patrick JE (1983) The autonomic nervous system and fetal heart rate variability. Am J Obstet Gynecol 146:456–462
8. Dawes GS, Houghton CRS, Redman CWG, Visser GHA (1982) Pattern of the normal human fetal heart rate. Br J Obstet Gynaecol 89:276–284
9. de Vries JIP, Visser GHA, Prechtl HFR (1982) The emergence of fetal behaviour. I. Qualitative aspects. Early Hum Dev 7:301–322
10. Leader LR, Baillie P, Martin B, Molteno C, Wynchank S (1984) Fetal responses to vibrotactile stimulation, a possible predictor of fetal and neonatal outcome. Aust N Z J Obstet Gynaecol 24:251–256
11. Nijhuis JG, Prechtl HFR, Martin CB, Bots RSG (1982) Are there behavioural states in the human fetus? Early Hum Dev 6:47–65
12. Patrick J, Campbell K, Carmichael L, Natale R, Richardson B (1982) Patterns of gross fetal body movements over 24-hour observation intervals during the last 10 weeks of pregnancy. Am J Obstet Gynecol 142:363–371
13. DiPietro JA, Caulfield LE, Costigan KA, Merialdi M, Nguyen RHN, Zavaleta N et al (2004) Fetal neurobehavioral development: a tale of two cities. Dev Psychol 40:445–456
14. Nijhuis I, ten Hof J, Mulder E, Nijhuis J, Narayan H, Taylor D et al (1998) Numerical fetal heart rate analysis: nomograms, minimal duration of recording, and intrafetal consistency. Prenat Neonatal Med 3:314–322
15. ten Hof J, Nijhuis IJM, Mulder EJH, Nijhuis JG, Narayan H, Taylor DJ et al (2002) Longitudinal study of fetal body movements: nomograms, intrafetal consistency and relationship with episodes of heart rate patterns A and B. Pediatr Res 52:568–575
16. Krasnegor N, Fifer W, Maulik D, McNellis D, Romero R, Smotherman W (1998) Fetal behavioral development: a transdisciplinary perspective for assessing fetal well-being and predicting outcome. Prenat Neonatal Med 3:185–190
17. Amiel-Tison C, Gosselin J, Kurjak A (2006) Neurosonography in the second half of fetal life: a neonatologist's point of view. J Perinat Med 34:437–446
18. DiPietro JA, Irizarry RA, Hawkins M, Costigan KA, Pressman EK (2001) Cross-correlation of fetal cardiac and somatic activity as an indicator of antenatal neural development. Am J Obstet Gynecol 185:1421–1428
19. Hepper PG (1995) Fetal behavior and neural functioning. In: Lecanuet JP, Fifer WP, Krasnegor NA, Smotherman WP (eds) Fetal development: a psychobiological perspective. Lawrence Erlbaum, Hillsdale, NJ, pp 405–417
20. Nijhuis IJM, ten Hof J (1999) Development of fetal heart rate and behavior: indirect measures to assess the fetal nervous system. Eur J Obstet Gynecol Reprod Biol 87:1–2
21. Sandman CA, Wadhwa P, Hetrick W, Porto M, Peeke H (1997) Human fetal heart rate dishabituation between thirty and thirty-two weeks gestation. Child Dev 68:1031–1040

22. Hepper PG, Shahidullah S (1992) Habituation in normal and Down's syndrome fetuses. Q J Exp Psychol 44:305–317
23. Horimoto N, Koyangi T, Maeda H, Satoh S, Takashima T, Minami T et al (1993) Can brain impairment be detected by in utero behavioural patterns? Arch Dis Child 69:3–8
24. Romanini C, Rizzo G (1995) Fetal behaviour in normal and compromised fetuses: an overview. Early Hum Dev 43:117–131
25. Maeda K, Morokuma S, Yoshida S, Ito T, Pooh R, Serizawa M (2006) Fetal behavior analyzed by ultrasonic actocardiogram in cases with central nervous system lesions. J Perinat Med 34:398–403
26. Nijhuis IJM, ten Hof J, Mulder EJ, Nijhuis JG, Narayan H, Taylor D et al (2000) Fetal heart rate in relation to its variation in normal and growth retarded fetuses. Eur J Obstet Gynecol Reprod Biol 89:27–33
27. Kainer F, Prechtl H, Engele H, Einspieler C (1997) Assessment of the quality of general movements in fetuses and infants of women with type-I diabetes mellitus. Early Hum Dev 50:13–25
28. Gingras JL, O'Donnell KJ (1998) State control in the substance-exposed fetus: I. The fetal neurobehavioral profile: an assessment of fetal state, arousal, and regulation competency. Ann N Y Acad Sci 846:262–276
29. Mulder EJ, Morssink LP, van der Schee T, Visser GH (1998) Acute maternal alcohol consumption disrupts behavioral state organization in the near term fetus. Pediatr Res 44:774–779
30. DiPietro J, Hodgson DM, Costigan KA, Johnson TRB (1996) Fetal antecedents of infant temperament. Child Dev 67:2568–2583
31. DiPietro JA, Costigan KA, Pressman EK, Doussard-Roosevelt J (2000) Antenatal origins of individual differences in heart rate. Dev Psychobiol 37:221–228
32. Lewis M, Wilson C, Ban P, Baumel M (1970) An exploratory study of resting cardiac rate and variability from the last trimester of prenatal life through the first year of postnatal life. Child Dev 41:799–811
33. Thomas PW, Haslum MN, MacGillivray I, Golding MJ (1989) Does fetal heart rate predict subsequent heart rate in childhood? Early Hum Dev 19:147–152
34. Groome L, Swiber M, Holland S, Bentz L, Atterbury J, Trimm R (1999) Spontaneous motor activity in the perinatal infant before and after birth: stability in individual differences. Dev Psychobiol 35:15–24
35. DiPietro JA, Bornstein MH, Costigan KA, Pressman EK, Hahn CS, Painter K et al (2002) What does fetal movement predict about behavior during the first two years of life? Dev Psychobiol 40:358–371
36. Degani S, Leibovitz Z, Shapiro I, Ohel G (2009) Twins' temperament: early prenatal sonographic assessment and postnatal correlation. J Perinatol 29:337–342
37. DiPietro JA, Costigan KA, Pressman EK (2002) Fetal state concordance predicts infant state regulation. Early Hum Dev 68:1–13
38. Groome L, Singh K, Bentz L, Holland S, Atterbury J, Swiber M et al (1997) Temporal stability in the distribution of behavioral states for individual human fetuses. Early Hum Dev 48:187–197
39. Richards T, Newbery H (1938) Studies in fetal behavior: III. Can performance on test items at six months postnatally be predicted on the basis of fetal activity? Child Dev 9:79–86
40. DiPietro J, Kivlighan K, Costigan K, Rubin S, Shiffler D, Henderson J et al (2010) Prenatal antecedents of newborn neurological maturation. Child Dev 81(1):115–130
41. Gualtney J, Gingras J (2005) Fetal rate of behavioral inhibition and preference for novelty during infancy. Early Hum Dev 81:379–386
42. Werner E, Myers M, Fifer W, Cheng B, Fang Y, Allen R et al (2007) Prenatal predictors of infant temperament. Dev Psychobiol 49:474–484
43. DiPietro JA, Ghera MM, Costigan KA (2008) Prenatal origins of temperamental reactivity in infancy. Early Hum Dev 84:569–575

44. Snidman N, Kagan J, Riordan L, Shannon D (1995) Cardiac function and behavioral reactivity during infancy. Psychophysiology 32:199–207
45. DiPietro JA, Bornstein MH, Hahn CS, Costigan KA, Achy-Brou A (2007) Fetal heart rate and variability: stability and prediction to developmental outcomes in early childhood. Child Dev 78:1788–1798
46. Bornstein MH, DiPietro JA, Hahn CS, Painter K, Haynes OM, Costigan KA (2002) Prenatal cardiac function and postnatal cognitive development: an exploratory study. Infancy 3:475–494
47. Ratcliffe S, Leader L, Heller G (2002) Functional data analysis with application to periodically stimulated foetal heart rate data. I: functional regression. Stat Med 21:1103–1114

Chapter 3
Maternal Influences on the Developing Fetus

Janet A. DiPietro

Keywords Pregnancy • Fetal movement • Fetal heart rate • Maternal anxiety/stress • Maternal–fetal interaction

> *"For behold, the moment that the sound of thy greeting came to my ears, the babe in my womb leapt for joy." Luke 1:44*

Observations of a link between pregnant woman and fetus, and speculation on its nature abound throughout history, literature, and across cultures. Despite the ubiquity of the phenomenon, relatively little is known about the manner in which the development of the fetus is influenced by the maternal psychological context. Scientific inquiry into the nature of this relationship has been historically hampered not by lack of interest, but by lack of access to the fetus. Although this changed considerably in the 1980s with the development of real time obstetric ultrasound, the prenatal period is the only time in development when interaction between mother and offspring cannot be directly observed and evaluated.

As discussed earlier in this volume, the scientific focus of fetal neurobehavioral assessment has been typically directed at the fetus, and not at the maternal–fetal pair. However, as early as the 1930s, the Fels Longitudinal Study included the first systematic scientific inquiry into factors that influence neurobehavioral functioning of the human fetus. Potential fetal influences considered included exposures such as cigarette smoking and nutritional factors as well as maternal psychological factors of emotionality and stress [1, 2]. Although the methods of access to the fetus and measurement of fetal heart rate and motor behavior were quite rudimentary, many of the questions posed then remain now as active areas of research. The preliminary findings from this project could have served to provide a broad substrate for a new field of scientific study. However, research into whether and how

J.A. DiPietro (✉)
Department of Population, Family and Reproductive Health, Johns Hopkins University, 615 N. Wolfe St., E4531, Baltimore, MD 21205, USA
e-mail: jdipietr@jhsph.edu

A.W. Zimmerman and S.L. Connors (eds.), *Maternal Influences on Fetal Neurodevelopment: Clinical and Research Aspects*,
DOI 10.1007/978-1-60327-921-5_3, © Springer Science+Business Media, LLC 2010

normal variation in maternal psychological state may influence the fetus was not intensively pursued in depth until much later in the century when, presumably, ultrasound technology facilitated access to the fetus and scientific inquiry.

Since the early 1990s, the Johns Hopkins Fetal Neurodevelopment Project has been collecting information focused on documenting normal human development and the factors that affect it. Our initial focus was on assessment of the developing fetus, much in the same way one would approach infant assessment. However, a number of anecdotal observations during the course of early data collection, as well as descriptions of maternal–fetal interactions by participants, changed this. These included evident accelerations of the audible signal generated by the fetal heart rate monitor when pregnant women related distressing stories and numerous anecdotes by participants to the effect of "I know that when I do 'X,' the fetus will do 'Y'." As a result, we reoriented our focus away from the fetus alone to the maternal–fetal dyad. Because there are no direct neural or circulatory connections between the pregnant woman and fetus, any fetal response to maternal psychological state or other experience that does not impinge mechanically on the uterine environment (e.g., maternal change in posture) requires signal transduction. As a result, a data collection system was developed to collect maternal psychophysiological data simultaneously with fetal neurobehavioral data. Maternal data collection includes heart period, a measure derived from a three-lead electrocardiogram, and respiration; in combination these provide a measure of respiratory sinus arrhythmia, which is an indicator of parasympathetic tone [3]. In addition, electrodermal activity, which reflects changes in conductivity of the skin, is measured by two electrodes applied to the fingertips. Skin conductance is mediated by the eccrine glands, which are singularly innervated by the sympathetic branch of the autonomic nervous system [4]. Fetal neurobehavioral data are collected using an actocardiograph, a fetal monitor that uses Doppler technology to detect fetal heart rate and motor activity. Maternal and fetal data are simultaneously digitized at 1,000 Hz, using streaming software on a PC-based system that permits analysis of each in relation to one another and to experimental manipulations. The routine monitoring period is 50 min in length. A schematic of this system is presented in Fig. 3.1. This chapter will review the information collected to date by the Johns Hopkins Fetal Neurobehavioral Development project that links the functioning of the maternal–fetal pair and, when possible, embeds it within the context of other current research findings.

Contemporaneous Associations Between Maternal Psychological Functioning and Fetal Neurobehavior

There has been significant interest in the role of maternal stress experienced during pregnancy in relation to outcomes ranging from preterm birth [5] to child development [6]. Although this subject will be returned to later in this chapter, the existing literature in this area stimulated our interest in documenting the potential effects of maternal stress in situ in an effort to understand potential mechanisms. The rationale

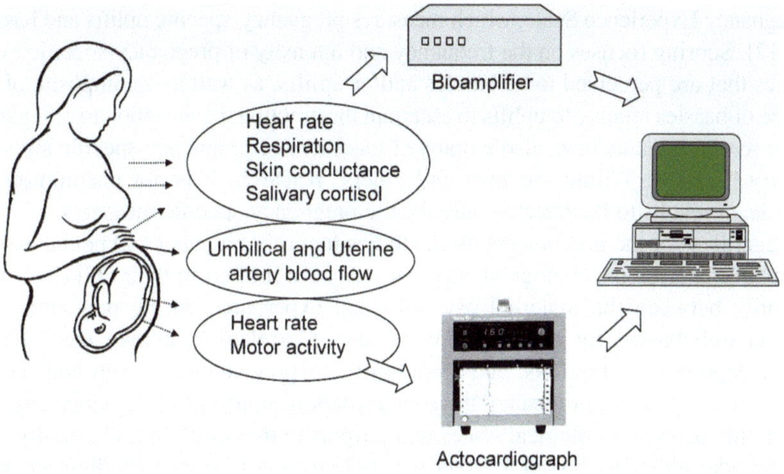

Fig. 3.1 Sources of data from a typical maternal–fetal protocol in the Johns Hopkins Fetal Neurobehavioral Project (Reprinted with permission: DiPietro (2005) Neurobehavioral assessment before birth. Ment Retard Dev Disabil Res Rev. 11:4–13, Wiley)

for doing so was that if there are persistent perinatal or postnatal developmental effects of maternal prenatal psychological functioning, there should be evidence for alterations to fetal development while the mediating biological influence was active. In a longitudinal study conducted at 24, 30, and 36 weeks of gestation, fetuses of women ($n=52$) who appraised their lives as more stressful and reported more frequent stresses or "hassles" that were specific to pregnancy were more motorically active [7]. However, fetal motor activity was also elevated in women who regarded themselves as more affectively intense and expressed greater negative vs. positive emotional valence toward pregnancy. These findings support another report of increased fetal motor activity in more anxious women [8]. More recently, the link between increased levels of pregnancy-specific stress and higher fetal motor activity was also detected from 24 to 38 weeks of gestation on a larger sample ($n=112$) [9], and in this case, variability in fetal heart rate was also higher. Although the tendency might be to conclude that the increased activity associated with greater maternal stress is an ominous outcome, higher levels of fetal motor activity were significantly predictive of more *optimal* motor and reflex maturation in the first few weeks of life [9]. This is consistent with the results generated by the Fels Study, which noted that greater fetal motor activity was predictive of advanced motor development at 6 months of age [10].

These and other findings described later in this chapter highlight the importance of measuring aspects of the maternal psychological experience that are most relevant to study participants. We have found that what women are most emotionally invested in during pregnancy, from both a positive and negative standpoint, is pregnancy. As unremarkable as this observation might seem on its surface, failure to measure pregnancy-related psychological factors will result in under-ascertainment of the maternal psychological milieu. As a result, we developed long and brief versions of a Pregnancy Experience Scale, which measures pregnancy-specific uplifts and hassles [11, 12]. Scoring focuses on the frequency and intensity of pregnancy-specific experiences that are perceived to be hassles and/or uplifts, as well as a composite of the degree of hassles relative to uplifts to ascertain the emotional connection to pregnancy. Other research teams have also employed measures of pregnancy-specific stress or anxiety [13, 14]. Within our own and others' research, it is not uncommon for significant results to be detected only for the pregnancy-specific measures.

Regardless of the instruments used, studies that rely on paper and pencil assessments of maternal psychological states or traits are limited in their ability to infer causality between the maternal psychological experience during pregnancy and fetal neurobehavior for a number of reasons. Correlations among measures of stress, depressive symptoms, and anxiety during pregnancy are high and remain relatively stable from pregnancy through 2 years postpartum [15], suggesting that commonly used psychological scales that purport to measure "stress" actually contain broader elements linked to other psychological attributes, including temperamental traits. These characteristics are known to affect maternal caregiving behavior in the postnatal period. Thus, postnatal measurement and control for these attributes is necessary to separate prenatal from postnatal influences on child outcomes.

The range of maternal psychological measures that are associated with fetal neurobehavior, and their persistence over time once pregnancy is over, raise the possibility that measurement of "prenatal stress" is neither specific to pregnancy nor specific to stress. Thus, in the studies described earlier, there is the possibility that both fetal motor activity and maternal psychological attributes reflect a shared genetic contribution without a causal relation. A recent investigation of the influence of a prenatal factor (i.e., maternal cigarette smoking) on child outcomes (i.e., child antisocial behavior) illustrates this issue. Comparisons between offspring conceived with in vitro fertilization using embryos from the mother and those conceived with donor embryos revealed associations between prenatal smoking and child behaviors only in the former [16]. This suggests that when child behavioral outcomes are linked to maternal smoking, it is less a result of biological effects of exposure to smoking on the developing fetus, and more, or perhaps only, a reflection of shared inheritance with characteristics of women who choose to smoke during pregnancy. It is not unlikely that a similar case could be made for findings linking prenatal maternal affect or stress to postnatal child behavior. Finally, the putative mechanism inferred in studies of stress and pregnancy, activation of the hypothalamic–pituitary–adrenal (HPA) axis, is of questionable validity since psychological assessments provided by nonclinical samples of women are largely unrelated to levels of HPA by-products, including cortisol [13, 14, 17, 18].

Fetal Response to Experimental Manipulation of Maternal State

A more effective, but methodologically challenging, way to evaluate whether the maternal psychological state affects the developing fetus is to manipulate maternal state and observe whether there is a fetal response. This approach provides both a temporal link between the dependent and independent measures and greater potential for revealing mechanisms. In the 1960s, investigators tried a variety of manipulations designed to alarm pregnant women and typically observed fetal tachycardia as a response [19]. We relied on a less threatening procedure, the Stroop Color–Word test, which is a challenging cognitive–perceptual task. Others have also employed this task in studies designed to evaluate fetal responsiveness to induced maternal stress [20, 21]. As expected, the procedure generated a clear physiological response in our sample of pregnant participants ($n = 137$). Fetuses responded to the manipulation, which lasted for about 4 min, with increased variability in heart rate and suppression of motor activity [22]. Once the manipulation was over, these values reverted to baseline levels, although male fetuses showed greater rebound in motor activity. There was moderate stability over time in the degree of both the maternal physiological response and the degree of fetal reactivity, suggesting that individual mothers and fetuses have their own characteristic response patterns.

The second manipulation used that was designed to activate maternal arousal was a labor and delivery documentary shown at 32 weeks of gestation to women who also participated in the Stroop study. The stimulus film included women relating their birth experiences interspersed with labor and delivery scenes. Again, maternal physiological data confirmed the effectiveness of the manipulation and again, fetuses responded with decreased motor activity, but in contrast to the Stroop intervention, decreased heart rate variability [23]. However, examination of the fetal response to a specific component of the documentary – the first graphic birth scene – revealed a somewhat different pattern of responsiveness. Fetuses of women who had not given birth before showed a transient increase in motor activity during this scene. This suggests that maternal influences on fetal neurobehavior may have biphasic elements and generate both acute (i.e., rapid and transitory) and more tonic (i.e., persistent and incremental) effects. This could also serve to reconcile the observation of higher levels of fetal motor activity linked to psychological attributes as described in the previous section with those reported here.

Our third manipulation was a relaxation procedure designed to examine how the fetal response to maternal arousal reduction may differ from arousal induction as effected through the Stroop task and labor and delivery film. A guided-imagery audiotape was used to induce an 18-min period of relaxation in pregnant women ($n = 100$) at 32 weeks of gestation. The manipulation generated the expected reduction in maternal psychological and physiological tension based on self-report measures of relaxation and physiological measurement of maternal respiration, heart rate, and skin conductance. The fetal response included decreased heart rate and increased heart rate variability but attributing these to the relaxation procedure itself could not be distinguished from simple maternal rest. However, there was a clear suppression in fetal motor activity during the manipulation, which recovered after the relaxation protocol concluded [24].

Although these studies have successfully demonstrated that the fetus responds to induced changes in maternal psychological state, they have been relatively unsuccessful in determining how this happens. In each study, change scores were computed for maternal physiological and fetal neurobehavioral responses from baseline to the manipulation (i.e., reactivity), and from the manipulation to the postmanipulation periods (i.e., recovery). However, significant associations between the degree of maternal and fetal responsiveness were not detected in all cases, and when they were detected, the magnitude of the association was typically modest, a phenomenon reported elsewhere [21]. In general, associations between maternal heart rate and fetal measures are weakest; those with maternal skin conductance somewhat stronger. However, the intrauterine environment is affected by more than the predominantly autonomic indicators measured by our data collection system. Concurrent maternal salivary cortisol and umbilical blood flow data collected in the relaxation study provided an opportunity to examine the potential role of HPA activation and oxygen transfer in understanding the observed fetal response in that protocol. However, detected associations were also few and modest. A decline in maternal salivary cortisol was significantly associated with the decline in fetal movement observed from baseline to the relaxation period, and the increase in fetal heart rate variability was associated with the degree of decline in umbilical artery resistance [24]. However, the degree of shared variance accounted for by these associations was 9% or less. Thus, understanding of the manner in which maternal experiences are transduced to the fetus remains incomplete.

If the relaxation manipulation had been implemented prior to the two stressful procedures, it would have been tempting to speculate that the fetus "relaxed" when the mother relaxed. Rather, because motor activity suppression is the same pattern of fetal response noted to other maternal manipulations, we offer another possibility: that the observed fetal responses in all of the studies were elicited by fetal alerting to sensory-based alterations in the intrauterine milieu. Fetal heart rate responses have been observed within seconds of maternal manipulations that impinge on the intrauterine environment, including maternal postural changes [25] and auditory stimuli [26]. A similarly rapid onset of a fetal response to induced maternal psychological stress, including increased fetal heart rate variability, has been reported in nonhuman primates [27].

Studies based on recordings conducted from within an intact uterus in animal preparations have found that the uterus is a fairly noisy place, with prominent maternal vasculature, digestive, and vocal sounds [28]. We suggest that sudden accelerations or decelerations in maternal heart rate and accompanying blood pressure and gastric motility changes that are elicited by variation in maternal psychological state provide the fetus with a changing uterine sensory environment. This may induce a fetal orienting response, consistent with the observation of suppressed cardiac variability and motor activity observed in the results reviewed here. Thus, women who have temperamentally more intense and volatile affect may present a different level of daily stimulation to fetuses throughout pregnancy than those who are more even-tempered. The results showing greater activity levels in fetuses of more affectively intense women that appraise their lives as stressful may indicate a biphasic or rebound effect of such lability on fetal neurobehavior.

Temporal Associations Between Maternal and Fetal Functioning

The next series of studies was targeted at investigating whether maternal and fetal measures are temporally related during undisturbed conditions. Maternal and fetal heart rate are correlated when averaged over periods of time ranging from 50 min [29] to 24 h [30] such that women with faster heart rates have fetuses with faster heart rates. However, an association such as this does not imply that one drives the other. To examine the nature of the temporal relationship, time series analysis was applied to continuous streams of two maternal measures, heart rate and skin conductance, and two fetal measures, heart rate and motor activity, collected during an undisturbed monitoring period of 50 min. Data were analyzed for 137 maternal–fetal pairs measured longitudinally at 20, 24, 28, 32, 36, and 38 weeks of gestation. Cross-correlation coefficients were computed for each cross-lagged value ranging from 0 s (i.e., coincident) to ±100 s. Results indicated that maternal and fetal heart rate were unrelated; that is, there were no systematic relations between change in one and change in the other. Therefore, any time independent associations that may exist between maternal and fetal heart rate must be generated by secondary processes that mediate both. However, clear associations emerged for maternal skin conductance and fetal motor activity with a time lag of 2 s, that is, the association between changes in these measures peaked at a 2 s interval. The magnitude and shape of this cross-correlation function did not change during the period of gestation evaluated. However, the direction of the association surprised us: fetal movement preceded maternal skin conductance changes. Put most simply, fetal motor activity stimulated a small maternal sympathetic "jolt" 2 s after it occurred [29]. An equally consistent, but slightly lower in magnitude, association between fetal movement and maternal heart rate was also demonstrated.

Working with colleagues, these findings were subsequently replicated in a sample of 195 maternal–fetal pairs in Lima, Peru measured using the same methodology and longitudinal intervals. Cross-correlation results again yielded no relationship between maternal and fetal heart rate, but similar associations between maternal skin conductance and both fetal parameters, with the same temporal lag [31]. Figure 3.2 presents the associations between fetal motor activity and maternal skin conductance in both samples of maternal–fetal pairs. Women generally detect only the largest and most sustained fetal movements [32], which means that the maternal response is evoked in the absence of perception most of the time. Since the fetus moves frequently, approximately once per minute or slightly more, during the second half of pregnancy [33–35], it would appear that pregnant women neither habituate nor become sensitized to this internal signal. In addition, stability was observed in the magnitude of these associations such that maternal–fetal pairs that show higher levels of synchrony at 20 weeks maintain higher levels through term.

The remarkable similarity in results generated from maternal–fetal pairs that are separated by geography, ethnicity, and socioeconomic prosperity suggests that such microanalytic techniques can provide information regarding fundamental properties

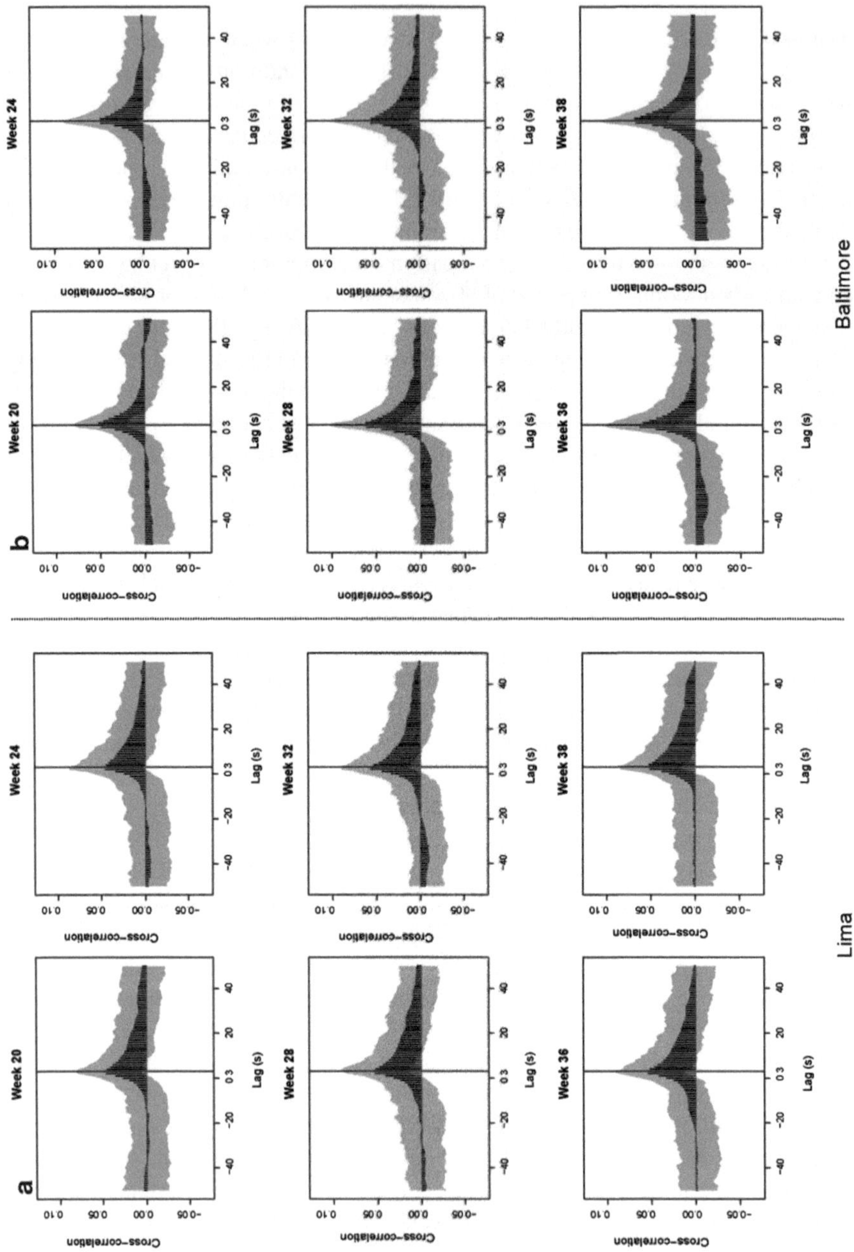
a
b
Week 20
Week 24
Week 28
Week 32
Week 36
Week 38
Cross-correlation
Lag (s)
Lima
Baltimore

of the synchrony within the maternal–fetal relationship. It is possible that fetal movement may generate an autonomic response mediated by adrenergic feedback mechanisms resulting from perturbations to the uterine wall. Although the mechanism is unknown, the results point to a role of the fetus in effecting a maternal response, indicating that the maternal–fetal relationship is bidirectional in nature.

A seminal paper by Bell [36] challenged the prevailing belief that parent–child interaction was a unidirectional phenomenon with the child as recipient of parental action. This led to the now well-accepted view of the maternal–child relationship as dynamic and transactional. A similar model has been applied in understanding the effects that rodent offspring have on the development of the maternal nervous system [37]. We have proposed [29] that sympathetic activation of the mother by the fetus may contribute to observed decrements in cognitive performance associated with advancing pregnancy [38, 39] by interfering with parasympathetic processes that are required for maintenance of attention. In turn, the coincident dampening of responsiveness to external physical and mental challenges [40–42] may allow heightening of responses to internal signals. Thus, periodic sympathetic surges generated by fetal activity may serve to entrain maternal arousal patterns to the behavior of the fetus with implications for the impending demands of newborn care.

Relationship Between Maternal Psychological State During Pregnancy and Postnatal Developmental Outcomes

As noted earlier, our original impetus for examining the maternal–fetal relationship was based on the reports that maternal psychological stress during pregnancy negatively affected infant and child development. To date, we have found no evidence to support this notion, and in two studies, we have found a somewhat facilitative effect of maternal distress during pregnancy on child outcomes. In the first, higher levels of maternal prenatal anxiety, nonspecific stress appraisal, and depressive symptoms were associated with better motor development (i.e., PDI scores) on the Bayley Scales of Infant Development when children ($n=94$) were 2 years old. Mental Development Index (MDI) scores were associated with higher prenatal anxiety and depressive symptoms [43]. Analyses controlled for postnatal maternal psychosocial values at 6 weeks and 2 years postpartum. The only deleterious effect of prenatally measured attributes involved pregnancy-specific stress. Offspring of women who perceived pregnancy more negatively than positively showed somewhat poorer emotional regulation and attention.

Fig. 3.2 Cross-correlation functions for fetal movement–maternal skin conductance collected longitudinally at six gestational ages on maternal–fetal pairs in Baltimore and Lima. The *y*-axis depicts the strength of the temporal association at multiple points from ±40 s from the origin. The *x*-axis shows the strongest association between fetal movement and subsequent maternal skin conductance at +3 s. (Reprinted with permission: DiPietro et al. (2006) Prenatal development of intrafetal and maternal-fetal synchrony. Behav Neurosci 120:687–701, American Psychological Association)

Because developmental assessments such as the Bayley Scales do not include evaluation of information processing, a more recent study utilized brainstem auditory evoked potentials (BAEP) as an indicator of neural development. Results indicate that greater pregnancy-specific stress (but not nonspecific anxiety) averaged over the second half of pregnancy was negatively associated with transmission speed within components of the BAEP waveform [9]. Faster transmission indicates enhanced maturation of the neural components of the auditory pathway; thus a negative association indicates that maternal stress had a maturative effect. There is a developing body of evidence that might be germane to these findings. That is, higher maternal cortisol after 31 weeks of gestation is associated with more advanced physical and neuromuscular neonatal maturation [44] and higher MDI scores when infants are 1-year old [14] but cortisol levels earlier in pregnancy show either no, or the opposite, association with outcomes. Thus timing of exposure to maternal cortisol levels during pregnancy may be a critical element.

However, we also found that maternal pregnancy-specific stress was positively associated with neonatal irritability as assessed during a neurological examination [9], a finding that partially supports other studies that have detected associations between maternal stress and/or anxiety and reduced infant behavioral regulation [45–47]. However, early infant irritability is a temperamental characteristic and not a phenomenon that portends impeded developmental outcomes. Thus, if there is a link between prenatal maternal psychological factors and early irritability, it should not be regarded as indicative of damage to the developing nervous system.

Our inability to detect deleterious effects of maternal psychological functioning on developmental outcomes or neurological development may seem to contradict the existing literature. However, examination of a number of widely cited studies purporting to show damaging effects of maternal stress, anxiety, and/or depression on infant outcomes reveal a number of significant methodological and interpretative problems. The most problematic has been the reliance on maternal report of child outcomes as opposed to measuring child outcomes by observational data collection. It is well-documented that maternal perception of child behavior and development is contaminated by maternal psychological characteristics in the direction of the reported findings. That is, women who are more anxious or stressed perceive their children to be more difficult, thereby inextricably confounding the dependent and independent measures in a direction that ensures detection of positive associations. Another common limitation is failure to measure and control for postnatal maternal psychological functioning since it is clear that women who are anxious or stressed during pregnancy remain so through at least the first 2 years of their child's life [15]. The postnatal influence on child-rearing as a function of maternal psychological status is also well-known; thus observed associations may be environmentally, as opposed to biologically, mediated.

Animal models provide the most compelling support for a link between prenatal stress and developmental deficits [48], but functional deficits are not uniformly found and some studies report benefits [49, 50]. Most importantly, animal models of stress rely on repeated and systematic experimental manipulations, which are very different from the nature of observational studies of psychological attributes of women. These issues have been discussed in depth elsewhere [51]. Our findings are in agreement with the long-standing viewpoint that the human brain requires sufficient, but not excessive, stress to promote neural development both before [52] and after birth [53]. Participants in our studies and similar studies in pregnancy based on volunteer samples tend to be fairly well-educated, financially stable women, who are free from significant psychopathology. Although they experience significant levels of stress and anxiety, it is likely to be qualitatively different from that experienced by impoverished or clinically treated groups of women. Thus, the upper range included in these studies might best be regarded as reflecting "moderate" stress as opposed to more toxic or debilitating exposures.

Summary and Future Directions

This review was not intended to be a comprehensive examination of the maternal influences on the fetus but rather to illustrate a progression of studies designed to reveal the complexities of the maternal–fetal relationship. There is perhaps no relationship in life as profound as the first one, and much of the understanding of how each member of the dyad influences the other awaits discovery. Since the degree of synchrony within individual maternal–fetal pairs shows evidence of being a stable characteristic of that pair, the next question is whether the degree of maternal–fetal synchrony presages the degree of postnatal synchrony in maternal–child interaction within given pairs. Are women who are more physiologically responsive to fetal movements more responsive to infant behavior? Does the degree to which the fetus responds to maternal changes in psychological state translate to an infant's success at serving as an elicitor of caregiving? Efforts to address these questions are currently underway in our laboratory using a maternal–infant physiological and behavioral interaction paradigm at 6 months postpartum. In any case, it is clear that just as neurodevelopment does not commence with birth, neither does the maternal–child relationship.

References

1. Sontag LW, Richards TW (1938) Studies in fetal behavior: I. Fetal heart rate as a behavioral indicator. Monogr Soc Res Child Dev 3(4 Serial No. 17):1–67
2. Sontag LW (1941) The significance of fetal environmental differences. Am J Obstet Gynecol 42:996–1003
3. Berntson G, Cacioppo J, Quigley K (1993) Respiratory sinus arrhythmia: autonomic origins, physiological mechanisms, and psychophysiological implications. Psychophysiology 30:183–196
4. Venables PH (1991) Autonomic activity. Ann N Y Acad Sci 620:191–207
5. Paarlberg KM, Vingerhoets A, Passchier J, Dekker G, van Geijn H (1995) Psychosocial factors and pregnancy outcome: a review with emphasis on methodological issues. J Psychosom Res 39:563–595

6. Van den Bergh B, Mulder E, Mennes M, Glover V (2005) Antenatal maternal anxiety and stress and the neurobehavioral development of the fetus and child: links and possible mechanisms. A review. Neurosci Biobehav Rev 29:237–258
7. DiPietro JA, Hilton SC, Hawkins M, Costigan KA, Pressman EK (2002) Maternal stress and affect influence fetal neurobehavioral development. Dev Psychol 38:659–668
8. Field T, Diego M, Hernandez-Reif M, Schanberg S, Kuhn C, Yando R, Bendell D (2003) Pregnancy anxiety and comorbid depression and anger: effects on the fetus and neonate. Depress Anxiety 17:140–151
9. DiPietro J, Kivlighan K, Costigan K, Rubin S, Shiffler D, Henderson J, Pillion J (2010) Prenatal antecedents of newborn neurological maturation. Child Dev 81:115–130
10. Richards T, Newbery H (1938) Studies in fetal behavior: III. Can performance on test items at six months postnatally be predicted on the basis of fetal activity? Child Dev 9:79–86
11. DiPietro JA, Ghera MM, Costigan KA, Hawkins M (2004) Measuring the ups and downs of pregnancy. J Psychosom Obstet Gynaecol 25:189–201
12. DiPietro JA, Christensen A, Costigan KA (2008) The pregnancy experience scale – brief version. J Psychosom Obstet Gynaecol 29:262–267
13. Buitelaar J, Huizink A, Mulder E, de Medina Robles P, Visser G (2003) Prenatal stress and cognitive development and temperament in infants. Neurobiol Aging 24:S53–S60
14. Davis E, Sandman C (2010) The timing of prenatal exposure to maternal cortisol and psychosocial stress is associated with human infant cognitive development Child Dev 81:131–148
15. DiPietro JA, Costigan KA, Sipsma H (2008) Continuity in self-report measures of maternal anxiety, stress, and depressive symptoms from pregnancy through two years postpartum. J Psychosom Obstet Gynaecol 29:115–124
16. Rice F, Harold G, Bolvin J, Hay D, van den Bree M, Thapar A (2009) Disentangling prenatal and inherited influences in humans with an experimental design. Proc Natl Acad Sci U S A 106:2464–2467
17. Gutteling B, de Weerth C, Zandbelt N, Mulder E, Visser G, Buitelaar J (2006) Does maternal prenatal stress adversely affect the child's learning and memory at age six? J Abnorm Child Psychol 34:789–798
18. Petraglia F, Hatch M, Lapinski R, Stomati M, Reis F, Cobellis L, Berkowitz G (2001) Lack of effect of psychosocial stress on maternal corticotropin-releasing factor and catecholamine levels at 28 weeks of gestation. J Soc Gynecol Investig 8:83–88
19. Copher DE, Huber C (1967) Heart rate response of the human fetus to induced maternal hypoxia. Am J Obstet Gynecol 98:320–335
20. Monk C, Myers MM, Sloan RP, Ellman LM, Fifer WP (2003) Effects of women's stress-elicited physiological activity and chronic anxiety on fetal heart rate. J Dev Behav Pediatr 24:32–38
21. Monk C, Sloan RP, Myers MM, Ellman L, Werner E, Jeon J, Tager F, Fifer WP (2004) Fetal heart rate reactivity differs by women's psychiatric status: an early marker for developmental risk? J Am Acad Child Adolesc Psychiatry 43:283–290
22. DiPietro J, Costigan K, Gurewitsch E (2003) Fetal response to induced maternal stress. Early Hum Dev 74:125–138
23. DiPietro JA, Ghera MM, Costigan KA (2008) Prenatal origins of temperamental reactivity in infancy. Early Hum Dev 84:569–575
24. DiPietro J, Costigan K, Nelson P, Gurewitsch E, Laudenslager M (2008) Maternal and fetal responses to induced relaxation during pregnancy. Biol Psychol 77:11–19
25. Lecaneut JP, Jacquet AY (2002) Fetal responsiveness to maternal passive swinging in low heart rate variability state: effects of stimulation direction and duration. Dev Psychobiol 40:57–67
26. Groome L, Mooney D, Holland S, Smith L, Atterbury J, Dykman R (1999) Behavioral state affects heart rate response to low-intensity sound in human fetuses. Early Hum Dev 54:39–54
27. Novak MFS (2004) Fetal–maternal interactions: prenatal psychobiological precursors to adaptive infant development. Curr Top Dev Biol 59:37–60

28. Querleu D, Renard X, Boutteville C, Crepin G (1989) Hearing by the human fetus? Semin Perinatol 13:409–420
29. DiPietro JA, Irizarry RA, Costigan KA, Gurewitsch ED (2004) The psychophysiology of the maternal–fetal relationship. Psychophysiology 41:510–520
30. Patrick J, Campbell K, Carmichael L, Probert C (1982) Influence of maternal heart rate and gross fetal body movements on the daily pattern of fetal heart rate near term. Am J Obstet Gynecol 144:533–538
31. DiPietro J, Caulfield LE, Irizarry RA, Chen P, Merialdi M, Zavaleta N (2006) Prenatal development of intrafetal and maternal–fetal synchrony. Behav Neurosci 120:687–701
32. Johnson TRB, Jordan ET, Paine LL (1990) Doppler recordings of fetal movement: II. Comparison with maternal perception. Obstet Gynecol 76:42–43
33. DiPietro JA, Costigan KA, Shupe AK, Pressman EK, Johnson TRB (1998) Fetal neurobehavioral development: associations with socioeconomic class and fetal sex. Dev Psychobiol 33:79–91
34. Nasello-Paterson C, Natale R, Connors G (1988) Ultrasonic evaluation of fetal body movements over twenty-four hours in the human fetus at twenty-four to twenty-eight weeks' of gestation. Am J Obstet Gynecol 158:312–316
35. ten Hof J, Nijhuis IJM, Mulder EJH, Nijhuis JG, Narayan H, Taylor DJ, Visser GHA (1999) Quantitative analysis of fetal generalized movements: methodological considerations. Early Hum Dev 56:57–73
36. Bell RQ (1968) A reinterpretation of the direction of effects in studies socialization. Psychol Rev 75:81–95
37. Kinsley C, Madonia L, Gifford G, Tureski K, Griffin G, Lowry C, Williams J, Collins J, McLearie H, Lambert K (1999) Motherhood improves learning and memory. Nature 402:137–138
38. Buckwalter J, Stanczyk F, McCleary C, Bluestein B, Buckwalter D, Randin K, Chang L, Goodwin T (1999) Pregnancy, the postpartum, and steroid hormones: effects on cognition and mood. Psychoneuroendocrinology 24:69–84
39. de Groot R, Adam J, Hornstra G (2003) Selective attention deficits during human pregnancy. Neurosci Lett 340:21–24
40. DiPietro J, Costigan KA, Gurewitsch ED (2005) Maternal physiological change during the second half of gestation. Biol Psychol 69:23–38
41. Kammerer M, Adams D, von Castelberg B, Glover V (2002) Pregnant women become insensitive to cold stress. BMC Pregnancy Childbirth 2:8
42. Matthews KA, Rodin J (1992) Pregnancy alters blood pressure responses to psychological and physical challenge. Psychophysiology 29:232–240
43. DiPietro JA, Novak MF, Costigan KA, Atella LD, Reusing SP (2006) Maternal psychological distress during pregnancy in relation to child development at age two. Child Dev 77:573–587
44. Ellman LM, Schetter CD, Hobel CJ, Chicz-DeMet A, Glynn LM, Sandman CA (2008) Timing of fetal exposure to stress hormones: effects on newborn physical and neuromuscular maturation. Dev Psychobiol 50:232–241
45. Davis E, Snidman N, Wadhwa P, Glynn L, Dunkel-Schetter C, Sandman C (2004) Prenatal maternal anxiety and depression predict negative behavioral reactivity in infancy. Infancy 6:319–331
46. Gutteling B, de Weerth C, Buitelaar J (2005) Prenatal stress and children's cortisol reaction to the first day of school. Psychoneuroendocrinology 30:541–549
47. Huizink A, de Medina Robles P, Mulder E, Visser G, Buitelaar J (2002) Psychological measures of prenatal stress as predictors of infant temperament. J Am Acad Child Adolesc Psychiatry 41:1078–1085
48. Weinstock M (2001) Alterations induced by gestational stress in brain morphology and behavior of the offspring. Prog Neurobiol 65:427–451

49. Fujioka T, Fujioka A, Tan N, Chowdhury G, Mouri H, Sakata Y, Nakamura S (2001) Mild prenatal stress enhances learning performance in the non-adopted rat offspring. Neuroscience 103:301–307
50. Meek L, Burda K, Paster E (2000) Effects of prenatal stress on development in mice: maturation and learning. Physiol Behav 71:543–549
51. DiPietro JA (2004) The role of prenatal maternal stress in child development. Curr Dir Psychol Sci 13:71–74
52. Amiel-Tison C, Pettigrew AG (1991) Adaptive changes in the developing brain during intrauterine stress. Brain Dev 13:67–76
53. Huether G (1998) Stress and the adaptive self-organization of neuronal connectivity during early childhood. Int J Neurosci 16:297–306

Chapter 4
Implications of Maternal Programming for Fetal Neurodevelopment

Laura M. Glynn

Keywords Pregnancy • Fetus • Maternal behavior • Maternal brain • Cortisol/glucocorticoids • Estrogen • Oxytocin

Programming the Maternal Brain

The transition to motherhood is arguably the most fundamental and profound stage in the lifespan of a female. The end result is an extensive transformation of the female, affecting her behaviors, emotions, and motivations. The magnitude of physiological change required to produce successful parturition cannot be underestimated. The dynamic process of pregnancy results in alterations in maternal anatomy, physiology, and metabolism, with each organ adapting differently [1, 2]. Among the changes is the growth and development of a new organ, the placenta, that has immune, endocrine, and vascular properties [3]. In part because of, and in addition to, the influences of the placenta, a pregnant woman experiences increases in blood volumes and cardiac output, hypercoagulation, insulin resistance, and a shift from a T-helper cell (Th)-1 to a Th-2 cytokine profile in the immune system [1, 4–6]. These changes are among those comprising the extensive transformation of maternal physiology necessary to maintain the pregnancy and to prepare the maternal brain for the challenges of motherhood.

A growing body of literature suggests a remarkable neural plasticity associated with reproductive experience. In 1971, Marian Diamond provided a striking example of such plasticity, showing that the cortical size of pregnant rats housed in impoverished conditions matched those of nonpregnant rats housed in enriched conditions [7]. For the first time, Diamond's work demonstrated that pregnancy

L.M. Glynn (✉)
Department of Psychiatry and Human Behavior, University of California, Irvine,
333 The City Blvd. W, Suite 1200, Orange, CA 92868, USA
and
Department of Psychology, Chapman University
e-mail: lglynn@uci.edu

A.W. Zimmerman and S.L. Connors (eds.), *Maternal Influences on Fetal Neurodevelopment: Clinical and Research Aspects*,
DOI 10.1007/978-1-60327-921-5_4, © Springer Science+Business Media, LLC 2010

reshapes the female brain. More recent work with rodent models has confirmed that pregnancy produces permanent changes in brain structure and function [8–13]. These changes are present in brain regions and behaviors involved in maternal caretaking, both directly (e.g., recognition of young and attachment) and indirectly (e.g., spatial memory and stress responsivity).

Alterations in Maternal Attachment and Care Giving

As would be expected, the earliest work examining the effects of pregnancy on the female involved the onset of classic maternal behavior. A strong body of work with animal models supports the notion that the immediate responsiveness of the female to her offspring at birth is a result of exposures to the hormones of pregnancy and parturition [14]. These hormonal effects can be mimicked by treating the ovariectomized virgin female with the hormones that determine the onset of these behaviors (e.g., estrogen and progesterone), and blocking these hormones will delay the onset of maternal behavior [15].

In humans, there are a small number of studies demonstrating that variations in prenatal levels of estrogen, cortisol, and oxytocin influence the quality of postpartum maternal care [16–18]; these relationships are discussed in detail below. Further, a recent study indicates a unique role for the process of parturition in the development of maternal responsiveness. A functional magnetic resonance imaging (fMRI) study examining the effects of mode of delivery on reactions of mothers to the sounds of their babies' cries [19] revealed that those women who had given birth vaginally exhibited greater activation in brain regions involved in regulation of empathy, arousal, motivation, and reward circuits compared with those who had delivered by planned cesarean section for convenience. One obvious limitation to this study that cannot be dismissed is the possibility that preexisting differences were present among women who chose to deliver vaginally versus those who delivered by planned cesarean. It is not a conceptual leap to imagine that mothers who are more attuned to their infants prenatally and in the postpartum period are more likely to favor a vaginal delivery. Nonetheless, these findings are consistent with the premise that the physiological events that accompany parturition do enhance maternal care in humans.

Alterations in Stress Responsiveness

The hypothesis that pregnancy influences behaviors not directly relevant to maternal behavior per se, led to investigations of the influence of reproduction on stress regulation and responsiveness. Both physiological and behavioral indices in animal models suggest that stress responses in pregnancy are altered. Pregnant mice, rats, and ewes show reduced fear and anxiety behavior in a variety of stressful situations compared to nonpregnant animals [20–22]. Neumann et al. [23] have shown that animal models of both physical and emotional stress result in reduced hypothalamic–pituitary–adrenal (HPA) axis responses (adrenocorticotropic hormone (ACTH) and corticosterone) in pregnant rats compared with virgins. In addition, their work indicates that the ACTH response to the administration of exogenous corticotropin-releasing hormone (CRH) is reduced in pregnant versus virgin animals.

Two studies in humans have directly assessed changes in psychological responses to stress during pregnancy. Both examined affective responses to major life events [24, 25]. In the first study, it was determined that the timing of a major earthquake during pregnancy was related to the magnitude of the stress response to that earthquake. Women who experienced the earthquake early in pregnancy rated it as more stressful than those who experienced it late. In a second study, responses to a wide variety of life events were examined including: job loss, problems in a romantic relationship, legal trouble, and threats of physical harm. Again, the data suggested that events occurring early in pregnancy were experienced as more stressful than those same events occurring later in pregnancy. Further, this effect was not limited to a few events: 14 out of the 18 events showed the same pattern with early events being rated as more stressful.

Studies that examine physiological responses mirror the psychological findings and also indicate that the stress response is dampened as a result of pregnancy in humans (See [26] for a detailed review). Administration of CRH stimulates the synthesis and release of ACTH, which in turn stimulates the release of cortisol in nonpregnant women, but does not produce detectable responses in women in the third trimester of pregnancy [27]. Similarly, cortisol responses to cold pressor challenge are absent in pregnant women, but present in nonpregnant women [28]. When presented with either a physical or psychological stressor at 21–23 weeks' gestation, pregnant women's blood pressure responses are reduced compared with those of nonpregnant women's [29] Heart rate responses to challenge also are decreased at 24 weeks' gestation compared with a nonpregnant state [30]. Similarly, compared with the nonpregnant state, catecholamine responses to challenge are diminished when examined during the third trimester of pregnancy [31]. Together, the current studies suggest that changes in the physiological stress response are present as early as late second trimester.

A common theme that arises in considering the effects of prenatal programming on the maternal brain concerns whether or not these alterations are merely epiphenomena or serve a functional purpose. For example, is the downregulated physiological stress response purely the result of changes in maternal physiology necessary to maintain gestation and effect parturition, or do they provide some protection for mother and fetus from the adverse effects of stress? The duration of human pregnancy is long and involves a significant investment on the part of the mother. Early in pregnancy, it may be advantageous for the mother to respond to and transmit the effects of environmental stress to the fetal/placental unit, thus increasing the probability of a pregnancy failure. As pregnancy progresses, it may become adaptive to maintain the pregnancy even in times of environmental hardship. It is plausible that as maternal investment increases with gestation, environmental sensitivity decreases in order to ensure that environmental stress is less likely to result in an adverse birth outcome, such as prematurity.

This hypothesis generates two specific predictions regarding the effects of stress exposure on adverse birth outcomes. First, at the group level it suggests that as pregnancy progresses the strength of association lessens between a stressor and length of gestation. This is due to the fact that as pregnancy advances, the probability that a stress exposure translates into a shortened gestation is diminished because both the physiological and the psychological responses to stress are dampened. Second, at the level of the individual, differences in the propensity to show a dampening in stress responsiveness during the prenatal period should predict length of gestation. That is, those women who remain relatively more sensitive to stress will be more likely to deliver prematurely, even in the absence of differences in stress exposures. Further, when these women are exposed to stressors, the consequences will be more profound for them than for the women who show a greater amount of dampening of stress responses. Empirical demonstrations are consistent with both of these premises. It has been shown that the impact of stress exposure during pregnancy depends upon timing and that early exposures may be more likely to produce a preterm birth [24, 32]. Further, women who do not show the expected decrease in reports of generalized stress and anxiety during pregnancy are at increased risk for preterm delivery (PTD) [33].

Alterations in Cognitive Performance

The influence of reproduction on cognitive function also has been addressed in a complementary line of research. Kinsely et al. [10] initially showed that pregnancy results in improved spatial learning and memory in female rats during the postpartum period. It also has been shown that pregnant rats exhibit enhanced working memory [34] and that pregnancy results in improved social learning [35] and increased speed of prey capture during the postpartum period [9].

Estimates of the percentage of women who report impaired cognitive function during pregnancy range from 48 to 81% [36, 37]. A small, but growing body of literature has moved beyond the use of self reports of cognitive function to assessment with objective measures. With few exceptions, the majority of studies indicate diminished cognitive functioning across a range of measures [36, 38–49]. A recent meta-analysis of the 17 studies published over the last decade confirmed this finding [50]. Results of the analysis indicated deficits in two components of memory during pregnancy: recall memory (both immediate and delayed) and the executive component of working memory. Effects were not detected for recognition memory, short-term memory, or implicit memory. Further, only decrements in recall memory persisted into the postpartum period.

Both rodents and humans show alterations in cognitive function as a result of pregnancy. However, there is one puzzling inconsistency when comparing the human findings to those of the animal models, and this concerns the direction of the effects. Overwhelming evidence from literature regarding rodents supports a role for enhanced memory during and after pregnancy. However, in the small body of published work in humans, diminished performance is observed. This rodent–human paradox could be simply due to species differences – it is true that vast differences in the physiology of pregnancy exist. However, it is also possible that when faced with different tasks than in the studies cited above, or those more directly relevant to the care of offspring such as multitasking, performance under stress, or sensitivity to infant cues, human mothers might exhibit enhanced performance, just as the rodent mothers do. Additional postpartum studies of maternal cognitive function are needed.

Alterations in Brain Structure and Function

The brain regions that regulate the development and maintenance of maternal behaviors are well characterized in animal studies and include: the medial preoptic area (mPOA), the cingulate cortex, the prefrontal and orbitofrontal cortices, the nucleus accumbens, the amydala, the lateral habenula, and the periaqueductal gray [51]. Not only do these areas initiate and regulate maternal behavior, but reproduction has also been shown to permanently alter their structure and function. For example, Keyser-Marcus et al. [8] have shown that in pregnancy, there is an increase in cell body size in the mPOA and also that there is an increase in the number of basal dendritic branches and dendritic length in this area. Similarly, Shingo et al. [52] have demonstrated that pregnancy enhances proliferation in the rat forebrain subventricular zone, stimulating olfactory neurogenesis, and that these effects persist at least until 4 weeks postpartum.

As can be expected from alterations in behavior, reproductive experience also affects brain regions involved in learning and memory, as well as those that regulate stress responses. It has been demonstrated recently that pregnancy increases dendritic spine density in the CA1 region of the hippocampus in rats [11, 12]. Further, Wartella and colleagues [22] have shown that pregnant rats showed less *c*-fos immunoreactivity in the CA3 region of the hippocampus and in the basolateral amydala after stress exposure; these changes were correlated with reductions in fear behavior. These differences were still present after weaning, indicating their persistence into the postpartum period.

In humans, estrogen alterations due to menopause and the menstrual cycle are associated with alterations in brain structure [53, 54]. These exposures which are of a relatively small magnitude, strongly suggest that in humans, as in rodents, the massive hormonal changes of pregnancy should be associated with altered brain morphology. Despite this, only one study has examined the influence of pregnancy on human brain structure. Oatridge et al. [55], using MRI in a small group of women, documented decreased total brain volumes over the course of pregnancy. Volumes were lowest just prior to parturition and showed detectable increases by 6-weeks postpartum; further increases were apparent from 6 weeks to 6 months.

Maternal Programming Effects Are Cumulative and Persistent

Animal models also have now provided evidence that the effects of parity and reproductive experience on stress responses and cognitive function are cumulative and persistent. Attenuated stress responsiveness and enhanced cognitive function accompanying pregnancy persist throughout the lifespan in rats [13, 56]. In addition, Gatewood et al. [13] have shown that at 24 months of age (the approximate equivalent of 60 years of age in humans), parous females show lower levels of immunoreactive amyloid precursor protein (a marker of neurodegeneration) than virgin animals. Further, it appears that the effect of parity on aging processes in the brain is cumulative: multiparous females had the lowest levels of the protein precursor and primiparous females exhibited levels between those of multiparous and nulliparous animals.

In humans, very little is known about the persisting or cumulative effects of parity. The only existing literature that addresses the long-term influences of parity on neurological function has examined cognitive function in postmenopausal women. Parous women are more likely to develop Alzheimer's disease (AD) [57] and an increasing number of births is predictive of earlier age at onset of AD [58]. Consistent with these findings, in a large cohort of postmenopausal women, those who were nulliparous exhibited better cognitive function over an average period of 12.8 years than those who were parous [59]. The human and rodent literatures agree in that parity appears to exert lasting and cumulative effects on neurological function. However, future research is needed to reconcile the observation that these lasting effects appear to be positive for rodent mothers and negative for human mothers.

The extent to which the long-term effects of reproductive experience on behavior and the brain in rodents can be attributed to pregnancy, independent of pup exposure and mothering, is not well established. There have been repeated demonstrations of alterations in cognitive performance and morphological changes in the maternal brain prior to pup exposure (i.e. during pregnancy; [8, 11, 34]). However, maternal exposure to pups also may contribute by itself to the enhanced foraging and spatial memory abilities associated with motherhood. Importantly, the effects of pup exposure appear to be enhanced if the maternal brain has been primed by the hormones of pregnancy [10, 60, 61]. That is, providing virgin females with a litter of pups to foster has a positive influence on memory and foraging effectiveness. However, the effects are relatively weak and inconsistent compared to the effects of exposure in parous females. Across these studies, it has been demonstrated that parous females with their pups outperform nulliparous females with a foster brood and both of those groups exhibit enhanced performance compared to nulliparous females without pup exposure.

Maternal Programming Mechanisms

Although the physiology of pregnancy and parturition differ across species, there is considerable agreement in literature examining both rodents and primates, about plausible hormonal mechanisms for the initiation and maintenance of maternal behavior, with the majority of the work focusing on estrogen, oxytocin, and glucocorticoids.

Estrogens

During human pregnancy, levels of estrogens rise precipitously [62]. By the 14th week of pregnancy, levels already are seven times those seen during the peak of the menstrual cycle, and in the final weeks of pregnancy, levels are 30 times greater. High levels of circulating estrogens regulate uterine blood flow, are necessary for the initiation of parturition [63–65], and are instrumental for the development of maternal behaviors.

Rodent work has consistently confirmed a role for estrogen in the regulation of maternal behavior. Systemic administration of estrogen blockers significantly delays the onset of maternal behavior in pregnant females [14, 66]. Estrogen implants inserted into the mPOA, the brain region primarily responsible for initiation and maintenance of maternal behavior [15, 67], induce maternal behavior in pregnancy-terminated and ovariectomized nonpregnant rats [68, 69]. Further, the facilitating effects of estrogen administration on maternal behavior are more rapid and larger in pregnant or pregnancy-terminated animals than in virgin females, suggesting that the hormonal milieu of pregnancy "primes" the brain for the estrogen exposures that occur at the end of gestation [14].

In nonhuman primates, elevated levels of estrogen during the prenatal period are associated with increased frequency of interaction with infants and reduced probability of infant rejection [70, 71]. In addition, exogenous administration of estradiol to ovariectomized female macaques and marmosets, in levels similar to those of late pregnancy, results in increased interactions with nonrelated infants [70, 72].

In the only published study that examines the effects of reproductive hormones on maternal behavior in humans, Fleming et al. [17] demonstrated that those women who show larger increases in estrogens from early to late pregnancy report increased feelings of attachment to their infants at 6-weeks postpartum. These data are consistent with the animal literature and suggest that estrogen exposures during the prenatal period are instrumental in the development of maternal behavior in humans.

Oxytocin

Oxytocin (OT), a small peptide consisting of nine amino acids, plays a central role in the modulation of social cognition and social behavior. The OT gene, located on chromosome 20, is composed of three exons which encode the oxytocin–neurophysin prohormone (Pro-OT/Np), from which OT is enzymatically cleaved. Oxytocin is synthesized in the magnocellular neurosecretory cells that are located in the supraoptic and paraventricular nuclei of the hypothalamus, and in the parvocellular neurons of the paraventricular nuclei. Projections from the magnocellular cells of the hypothalamus link to the posterior pituitary, from which OT is released into the circulation. Oxytocin is secreted within the central nervous system through projections from the parvocellular cells of the paraventricular nuclei to the limbic system and to the mid- and hind-brain nuclei. Central and peripheral release of OT can occur independently or in a coordinated manner [73]. At the end of gestation, OT is released peripherally from the posterior pituitary to trigger uterine contractions necessary for parturition [63]. Few longitudinal studies have assessed OT levels across pregnancy, but most of those have documented an increase across gestation [74–77]. During lactation, OT holds a critical regulatory role by contracting myoepithelial tissue in the breast to promote milk letdown [78].

In the female rodent, central administration of OT increases affiliative behavior and pair-bonding, and OT antagonists inhibit bond formation [79–82]. Peripheral administration of OT also has been shown to increase the probability of mating and influence partner preference in the female voles and gerbils [83–86]. The central role of OT in controlling maternal behavior was first confirmed by Pederson and Prange [87] who showed that intraventricular treatment with OT could rapidly induce maternal behavior in virgin rats. These findings have been extended by demonstrations that treatment with OT antagonists or lesions of OT-producing neurons will inhibit the initiation of maternal behavior [88, 89]. Further, evidence suggests that maintenance of maternal behavior such as pup-licking and nursing postures in rodent mothers can be increased by administration of OT and decreased by administration of OT antagonists [90, 91].

Studies of the influence of OT on maternal-offspring bonding in nonhuman primates also confirm its role. Intracerebral administration of OT increases approach and touching of infants in nulliparous macaques [92]. In common marmosets, peripheral administration of an OT antagonist reduces maternal interest in infant behavior [93]. In addition, among free-ranging macaques, higher levels of peripheral OT are associated with indicators of maternal warmth, including time spent nursing and grooming [94].

Consistent with animal models indicating that peripheral administration of OT exerts clear effects on affiliative behaviors [83, 84, 86], there is evidence that OT is associated with social and affiliative behaviors in humans. For example, variations in OT levels have been linked to physical and sexual contact, relationship quality, social support, adult attachment, trust, generosity, increased gaze toward the eye region of the face, memory for faces, the ability to read emotional states, and reduced psychological and physiological stress responsiveness [95–109]. Although there is some debate about the mechanisms through which peripheral OT might influence behavior, rapidly accumulating evidence does support such an association.

Two recent studies provide the first empirical demonstrations of the role of OT in the regulation of human mother–infant bonding. First, Levine et al. [77] found that those women who experienced larger increases in plasma OT from early to late pregnancy reported feeling closer and more attached to their fetuses. In a second study, plasma OT levels at 10 weeks of gestation and at 2 weeks postpartum were related to both behavioral and self-report measures of maternal care at 2 weeks postpartum [16]. Specifically, those women who exhibited higher OT levels displayed more infant-directed gaze, affectionate touch, positive affect, and "motherese" vocalizations. Further, higher levels of OT were associated with reports of attachment to the infant and to reports of infant checking behavior (the extent to which mothers check on their infants during both day and night).

Glucocorticoids

During gestation, cortisol, the primary glucocorticoid (GC) in humans, reaches levels consistent with those seen in Cushing's Syndrome and major melancholic depression. GCs are essential for the regulation of intrauterine homeostasis, and differentiation and maturation of vital organ systems in the fetus including the lungs, liver, and CNS [110–112]. In the presence of this state of pregnancy-induced hypercortisolism, the HPA response to challenge is altered. Specifically, across a range of species including primates, the HPA response to challenge during gestation is diminished [22, 26, 27]. Further, this diminished responsiveness is perpetuated by lactation [113–116], and some evidence suggests that even beyond the cessation of lactation, HPA-axis regulation is permanently altered in parous rodents and women [22, 117–120].

In nonhuman animals, GCs facilitate pair bonding [121,] and males of naturally biparental species, including humans, exhibit elevations in GCs before the birth of their offspring, a characteristic not found in males of nonparental species [122, 123]. Accumulating evidence also suggests a role for GCs in maternal behavior. Rodents that are adrenalectomized during pregnancy exhibit impaired pup retrieval and spend less time on the nest over pups and licking them [124, 125]. Further, corticosterone replacement in adrenalectomized animals restores the maternal behaviors [125]. Similarly, in baboons, higher levels of prenatal maternal cortisol late in gestation are associated with higher behavioral ratings of infant-directed affiliative behaviors [126].

In humans, elevated cortisol levels during the first 2 days postpartum have been linked to the mother's increased affectionate touch, enhanced attractiveness of infant odors to the mother, the ability to identify her own infant's odor, and the ability to discriminate between infant pain and hunger cries [18, 127, 128]. To date, no published study has examined prenatal cortisol levels across gestation in relation to maternal attachment or behavior. However, based on the nonhuman animal work, it seems likely that an association would be revealed with appropriate studies.

Emerging Possibilities for Fetal Participation in Prenatal Programming of Maternal Brain

It is becoming increasingly recognized that maternal signals shape the development of the fetus. However, it is not as widely acknowledged that this is only one side of a bidirectional relationship, specifically that fetal or placental signals also may shape the development of the maternal brain and behavior. It is possible that the fetus exerts these influences through endocrine, cellular, and behavioral routes.

Placental CRH

CRH is a 41-amino acid neuropeptide that is synthesized primarily in the paraventricular nucleus of the hypothalamus and has a major role in regulating pituitary–adrenal function and the physiological response to stress [129]. The maternal HPA axis is altered dramatically during human pregnancy because the *placenta* also expresses the genes for CRH. Placental CRH (pCRH) increases several hundred-fold as pregnancy advances and reaches levels in the maternal circulation at term observed only in the hypothalamic portal system during physiological stress [130]. In contrast to the inhibitory influence of maternal stress signals (e.g., cortisol) on expression of the CRH gene in the hypothalamus, maternal cortisol activates the promoter region of the gene in the placenta and stimulates CRH synthesis [131, 132].

As noted above, little is known about prenatal influences of GCs, estrogens, and OT on human maternal brain and behavior. However, even less is known about the possible influences of pCRH. In the nonpregnant state, CRH is believed to play a role in the etiology of depression. Depressed individuals have an increased number and hypersensitivity of CRH neurons in the paraventricular nucleus of the hypothalamus [133, 134]. Because of the dramatic increase in pCRH during pregnancy and the link between CRH and depression, it is possible that pCRH exposures may present a risk for postpartum depression (PPD). In a cohort of 100 women followed prospectively five times beginning early in pregnancy, elevations in midgestational pCRH were associated with an increased risk of developing symptoms of PPD. Specifically, pCRH levels at 25 weeks of gestation accurately identified 75% of women who subsequently would develop PPD symptoms [135]. These findings add new support to the small but emerging literature indicating that the maternal brain is susceptible to changes associated with normal human pregnancy and provides some of the first evidence that a fetal signal may reshape the maternal brain.

Fetal Sex

Fetal sex is an additional factor that has the potential to alter the prenatal endocrine milieu, and therefore, may have implications for maternal functioning. For example, in a longitudinal study of pregnant women, fetal sex predicted maternal memory performance [136]. Women who were carrying male fetuses showed better performance on spatial rotation and working memory tasks than their counterparts carrying female fetuses. These differences were apparent as early as 12 weeks of gestation, which makes it unlikely that these differences were due to production of sex steroids from the fetal gonads (production of testosterone by male fetuses does not peak until 15–18 weeks). The authors instead suggest that the differences may be due to levels of maternal serum human chorionic gonadotropin (hCG) levels which are dependent, in part, on fetal sex. hCG is a glycoprotein hormone produced by the developing embryo and later by the syncytiotrophoblasts (multinucleated placental cells responsible for the majority of the production of placental hormones). It readily enters the maternal circulation and traverses the blood–brain barrier. Specific receptors for hCG have been identified in many brain regions, including the hippocampus, which contains the highest density of receptors [137]. In pregnancies with female fetuses, maternal hCG levels are elevated compared to those with male fetuses, and this difference is apparent as early as 3 weeks after conception and persists throughout gestation [138–140].

Fetal Behavior

Endocrine signals do not represent the only pathway through which the fetus might shape the mother; an additional possibility is fetal behavior. In an examination of this link, DiPietro and colleagues [141] applied time series analysis to data from mother to fetus pairs six times during gestation, ranging from 20 to 38 weeks. They found consistent associations between fetal movement and maternal heart rate and skin conductance. Beginning at 20 weeks of gestation until term, fetal movement stimulated peak rises in maternal heart rate and skin conductance at 2 and 3 s after the event, respectively. Currently the pathway through which fetal movements might determine maternal sympathetic arousal is unknown. However, it is unlikely that this occurs through conscious perception of these movements. At term, women detect as few as 16% of fetal movements [142], which is consistent with the fact that although they are relatively skilled at detecting large or prolonged fetal movements, they are not very able to detect smaller spontaneous or evoked fetal movements [143]. Given that the pathway does not operate through conscious channels, DiPietro et al. [141] propose that the most likely local mechanism is through perturbations of the uterine wall. They go on to point out that these interactions may have broader implications for the role of the fetus in shaping maternal behavior. Specifically, they suggest that sympathetic activation in response to the fetal movement signal may begin to prepare the woman for new demands of motherhood by redirecting maternal resources away from competing, but less relevant environmental demands. As the authors further point out, this finding raises the additional provocative question of whether the degree of prenatal synchrony between mother and fetus might set the stage for postnatal mother–infant interaction.

Fetal Microchimerism

It is known that fetal cells cross the placental barrier and enter the maternal circulation, a process known as fetal microchimerism. This was first realized in 1979 when Herzenberg et al. [144] demonstrated the presence of cells containing a Y chromosome in the plasma of women who were pregnant with male fetuses. This finding was further expanded when similar male cells were demonstrated in the plasma of healthy women decades after giving birth to a son [145]. In the human, fetal cells have been detected in a range of maternal tissues including: thyroid, heart, liver, lungs adrenals, kidneys, and bone marrow [146, 147]. There is evidence that fetal cells are more prevalent in diseased tissues compared to healthy tissues, and this finding has inspired a debate as to whether microchimerism plays a role in triggering disease or the increase in fetal cells is associated with tissue repair [148]. More relevant to the issue of maternal programming are recent findings from a mouse model that has demonstrated the presence of fetal cells in the brain of pregnant mice [149]. This study indicated that these fetal cells were capable of taking on a range of attributes including neuron-, astrocyte-, and oligodendrocyte-like cell types. Further, there were more cells present at 4-weeks postpartum, than at parturition. Whether or not these cells have any functional or physiological significance has yet to be demonstrated. However, in the Tan et al. study, fetal cells were preferentially found in the region of the olfactory bulb. Previously, it has been shown that the hormonal changes of pregnancy stimulate neurogenesis in the subventricular zone of the forebrain in mice and that these neurons then migrate to the olfactory bulb to produce new interneurons [52]. Increases in olfactory interneurons have been linked to enhanced new odor memory [150], which plays a central role in offspring recognition in mammals. It is possible that pregnancy changes the attraction of specific brain areas for fetal cells.

Implications for the Study of Prenatal Influences on Neurodevelopment

Fetal Programming

Programming is a process by which a stimulus or insult during a critical developmental period has a long-lasting or permanent influence. Tissues grow and mature in a specific developmental sequence and different organs are sensitive to programming influences at different times depending upon their rate of cell division and further differentiation. Thus, the timing of the stimulus during development coupled with the time table for organogenesis and maturation determine the nature of the programmed effect. Fetuses exposed to maternal stress signals at various times during gestation are at subsequent risk for a range of adverse physical health outcomes including cardiovascular disease, hypertension, hyperlipidemia, insulin resistance, noninsulin-dependent diabetes mellitus, obesity, elevated serum cholesterol concentrations, and a shortened life span [151–155]. Exposure to maternal stress signals also has been linked to neurodevelopment and mental health with affected outcomes including internalizing and externalizing behavior, ADHD, deficits in cognitive performance and IQ, schizophrenia, and autism [156–158]. Therefore, the effects of fetal programming have important public health consequences related to multiple forms of morbidity.

Rationale for an Integrated Model of Fetal and Maternal Programming

In 1968, Bell [159] published a seminal paper that challenged existing models of the parent–child relationship as unidirectional, by suggesting that this relationship must be characterized as a bidirectional process occurring between parent and child. Just as this reciprocal relationship must be understood in the context of the parent–child relationship, similarly, in order to fully understand the persisting influences of the intrauterine environment on neurodevelopment, the effects of the prenatal environment on both the fetus and the mother, as well as their reciprocal influences, must be elucidated. This is critical, among other reasons, because the same hormones that program fetal development also are those that control and shape the maternal brain and behavior. Each of the hormones discussed above that plays a role in prenatal maternal programming also has been shown to influence fetal programming. Prenatal exposures to cortisol, OT, CRH, and estrogens have been shown to exert a wide range of influences on brain and behavior [160–164]. These findings in fetal programming, combined with those from the maternal programming literature, underscore the importance of a model of prenatal influences on neurodevelopment that includes maternal programming (see Fig. 4.1).

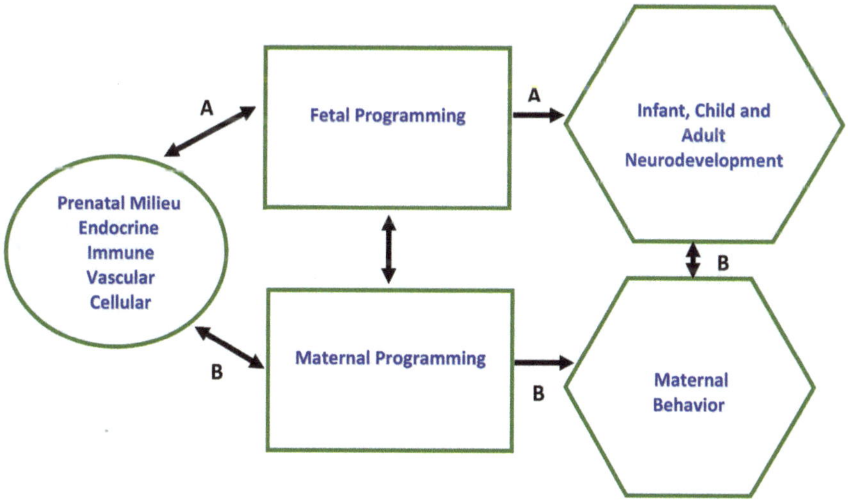

Fig. 4.1 An integrated model of maternal and fetal programming. The figure illustrates two potential routes through which the prenatal environment might influence neurodevelopment. The direct route (*Path A*) in which hormone exposures, for example, would shape fetal neurodevelopment, which in turn would be reflected in postnatal measures of neurodevelopment. The indirect route (*Path B*) involves prenatal influences, such as hormone exposures on the maternal brain and behavior, which then shape neurodevelopment in the postnatal period. These pathways are not mutually exclusive, and it is likely that both pathways contribute to prenatal influences on neurodevelopment

Perhaps the most obvious need for the integration of prenatal maternal programming into the study of fetal neurodevelopment involves the potential for maternal programming processes to account for or mediate apparent prenatal influences on fetal neurodevelopment. This can be illustrated by a consideration of the effects of hormone exposures on mother and fetus. Prenatal hormone exposures can shape neurodevelopment directly (Fig. 4.1, Path A). However, they also can produce the same effects on infant and child development indirectly, by shaping maternal behavior which then shapes infant and child neurodevelopment (Fig. 4.1, Path B). In addition to allowing the determination of the direct and indirect effects of prenatal hormone exposures on fetal neurodevelopment, the finding that fetal and maternal programming may occur in parallel raises interesting possibilities related to the adaptive significance of fetal programming and its long-term consequences. The human fetus may adjust its development in response to prenatal maternal stress signals, such as elevated pCRH, in anticipation of a hostile or nonnurturing postnatal environment. The fetus that is stressed in utero and adjusts its development accordingly in order to prepare for a hostile environment may cope better in the presence of lower quality of maternal care than the fetus that was not exposed to prenatal stress signals and did not make this anticipatory adjustment to its developmental trajectory. It is possible that prenatal signals that influence the postnatal maternal care and sensitivity serve to inform the fetus about the quality of the maternal environment it is likely to encounter.

References

1. Torgersen KL, Curran CA (2006) A systematic approach to the physiologic adaptations of pregnancy. Crit Care Nurs Q 29:2–19
2. Williams D (2003) Pregnancy: a stress test for life. Curr Opin Obstet Gynecol 15:465–471
3. Petraglia F, Florio P, Nappi C, Genazzani AR (1996) Peptide signaling in human placenta and membranes: autocrine, paracrine, and endocrine mechanisms. Endocr Rev 17:156–186
4. Wadhwa PD, Culhane JF, Rauh V et al (2001) Stress, infection and preterm birth: a biobehavioural perspective. Paediatr Perinat Epidemiol 15:17–29
5. Greer IA (1999) Thrombosis in pregnancy: maternal and fetal issues. Lancet 10:1258–1265
6. Kuhl C (1991) Insulin resistance in pregnancy and GDM: implications for diagnosis and management. Diabetes 40:18–24
7. Diamond MC, Johnson RE, Ingram C (1971) Brain plasticity induced by environment and pregnancy. Int J Neurosci 2:171–178
8. Keyser-Marcus L, Stafisso-Sandoz G, Geecke K et al (2001) Alterations of medial preoptic area neurons following pregnancy and pregnancy-like steroidal treatment in the rat. Brain Res Bull 55:737–745
9. Kinsely CH, Lambert KG (2006) The maternal brain. Sci Am 294:72–79
10. Kinsely CH, Madonia L, Gifford GW et al (1999) Motherhood improves learning and memory. Nature 402:137–138
11. Kinsely CH, Trainer R, Stafisso-Sandoz G et al (2006) Motherhood and the hormones of pregnancy modify concentrations of hippocampal neuronal dendritic spines. Horm Behav 49:131–142
12. Pawluski JL, Galea LAM (2006) Hippocampal morphology is differentially affected by reproductive experience in the mother. J Neurobiol 66:71–81

13. Gatewood JD, Morgan MD, Eaton M et al (2005) Motherhood mitigates aging-related decrements in learning and memory and positively affects brain aging in the rat. Brain Res Bull 66:91–98
14. Rosenblatt JS, Mayer AD, Giordano AL (1988) Hormonal basis during pregnancy for the onset of maternal behavior in the rat. Psychoneuroendocrinology 13:29–46
15. Numan M (2006) Hypothalamic neural circuits regulating maternal responsiveness toward infants. Behav Cogn Neurosci Rev 5:163–190
16. Feldman R, Weller A, Zagoory-Sharon O, Levine A (2007) Evidence for a neuroendocrinological foundation of human affiliation: plasma oxytocin levels across pregnancy and the postpartum period predict mother-infant bonding. Psychol Sci 18:965–970
17. Fleming AS, Ruble D, Krieger H, Wong PY (1997) Hormonal and experiential correlates of maternal responsiveness during pregnancy and the puerperium in human mothers. Horm Behav 31:145–158
18. Fleming AS, Steiner M, Corter C (1997) Cortisol, hedonics and maternal responsiveness. Horm Behav 32:85–98
19. Swain JE, Tasgin E, Mayes LC, Feldman R, Constable RT, Leckman JF (2008) Maternal brain response to own baby-cry is affected by cesarean section delivery. J Child Psychol Psychiatry 49:1042–1052
20. Maestripieri D, D'Amato FR (1991) Anxiety and maternal aggression in house mice (Mus musculus): a look at interindividual variability. Comp Psychol 105:295–301
21. Viérin M, Bouissou MF (2001) Pregnancy is associated with low fear reaction in ewes. Physiol Behav 72:579–587
22. Wartella J, Amory E, Macbeth AH et al (2003) Single or multiple reproductive experiences attenuate neurobehavioral stress and fear responses in the female rat. Physiol Behav 79:373–381
23. Neumann ID, Johnstone HA, Hatzinger M et al (1998) Attenuated neuroendocrine responses to emotional and physical stressors in pregnant rats involve adrenohypophysial changes. J Physiol 508:289–300
24. Glynn LM, Schetter CD, Wadhwa PD, Sandman CA (2004) Pregnancy affects appraisal of negative life events. J Psychosom Res 56:47–52
25. Glynn LM, Wadhwa PD, Dunkel-Schetter C, Chicz-DeMet A, Sandman CA (2001) When stress happens matters: effect of earthquake timing on stress responsivity in pregnancy. Am J Obstet Gynecol 184:637–642
26. de Weerth C, Buitelaar JK (2005) Physiological stress reactivity in human pregnancy – a review. Neurosci Biobehav Rev 29:295–312
27. Schulte HM, Weisner D, Allolio B (1990) The corticotropin releasing hormone test in late pregnancy: lack of an adrenocorticotropin and cortisol response. Clin Endocrinol 33:99–106
28. Kammerer M, Adams D, von Castelberg B, Glover V (2002) Pregnant women become insensitive to cold stress. BMC Pregnancy Childbirth 2:8, http://www.biomedcentral.com/1471-2393/2/8
29. Matthews KA, Rodin J (1992) Pregnancy alters blood pressure responses to psychological and physical challenge. Psychophysiology 25:41–49
30. DiPietro JA, Costigan KA, Gurewitsch ED (2005) Maternal psychophysiological change during the second half of gestation. Biol Psychol 69:23–38
31. Nisell H, Hjemdahl P, Linde B, Lunell NO (1985) Sypmpatho-adrenal and cardiovascular reactivity in pregnancy-induced hypertension. I. Responses to isometric exercise and a cold pressor test. Br J Obstet Gynaecol 92:554–560
32. Lederman SA, Rauh V, Weiss L et al (2004) The effects of the World Trade Center event on birth outcomes among term deliveries at three lower Manhatten hospitals. Environ Health Perspect 112:1772–1778
33. Glynn LM, Dunkel Schetter C, Hobel CJ, Sandman CA (2008) Pattern of perceived stress and anxiety in pregnancy predicts preterm birth. Health Psychol 27:43–51
34. Bodensteiner KJ, Cain P, Ray AS, Hamula LA (2006) Effects of pregnancy on spatial cognition in female Hooded Long-Evans rats. Horm Behav 49:303–314

35. Fleming AS, Kuchera C, Lee A, Winocur G (1994) Olfactory-based social learning varies as a function of parity in female rats. Psychobiology 22:37–43
36. Brindle PM, Brown MW, Brown J, Griffith HB, Turner GM (1991) Objective and subjective memory impairment in pregnancy. Psychol Med 21:647–653
37. Poser CM, Kassirer MR, Peyser JM (1986) Benign encephalopathy of pregnancy. Acta Neurol Scand 73:39–43
38. Buckwalter JG, Stanczyk FZ, McCleary CA et al (1999) Pregnancy, the postpartum, and steroid hormones: effects on cognition and mood. Psychoneuroendocrinology 24:69–84
39. Keenan PA, Yaldoo DT, Stress ME, Fuerst DR, Ginsberg KA (1998) Explicit memory in pregnant women. Am J Obstet Gynecol 179:731–737
40. Sharp K, Brindle PM, Brown MW, Turner GM (1993) Memory loss during pregnancy. Br J Obstet Gynaecol 100:209–215
41. Silber M, Almkvist O, Larsson B, Uvnäs-Moberg K (1990) Temporary peripartal impairment in memory and attention and its possible relation to oxytocin concentration. Life Sci 47:57–65
42. Janes C, Casey P, Huntsdale C, Angus G (1999) Memory in pregnancy. I: subjective experiences and objective assessment of implicit, explicit and working memory in primigravid and primiparous women. J Psychosom Obstet Gynaecol 20:80–87
43. Eidelman AI, Hoffman NW, Kaitz M (1993) Cognitive deficits in women after childbirth. Obstet Gynecol 81:764–767
44. Shetty DN, Pathak SS (2002) Correlation between plasma neurotransmitters and memory loss in pregnancy. J Reprod Med 47:494–496
45. de Groot RHM, Adam JJ, Hornstra G (2003) Selective attention deficits during human pregnancy. Neurosci Lett 340:21–24
46. de Groot RHM, Hornstra G, Roozendaal N, Jolles J (2003) Memory performance, but not information processing speed, may be reduced in early pregnancy. J Clin Exp Neuropsychol 25:482–488
47. Casey P, Huntsdale C, Angus G, Janes C (1999) Memory in pregnancy. II: implicit, incidental, explicit, semantic, short-term, working and prospective memory in primigravid, multigravid and post-partum women. J Psychosom Obstet Gynaecol 20:158–164
48. Crawley RA, Dennison K, Carter C (2003) Cognition in pregnancy and the first year postpartum. Psychol Psychother 76:69–84
49. McDowall J, Moriarty R (2000) Implicit and explicit memory in pregnant women: an analysis of data-driven and conceptually driven processes. Q J Exp Psychol 53A:729–740
50. Henry JD, Rendell PG (2007) A review of the impact of pregnancy on memory function. J Clin Exp Neuropsychol 29:793–803
51. Numan M (1994) Maternal behavior. In: Knobil E, Neill J, Ewing JLL, Greenwald GS, Markert CL, Pfaff DW (eds) The physiology of reproduction. Raven, New York, pp 221–302
52. Shingo T, Gregg C, Enwere E et al (2003) Pregnancy-stimulated neurogenesis in the adult female forebrain mediated by prolactin. Science 299:117–120
53. Lord C, Buss C, Lupien SJ, Pruessner JC (2008) Hippocampal volumes are larger in postmenopausal women using estrogen therapy compared to past users, never users and men: a possible window of opportunity effect. Neurobiol Aging 29:95–101
54. Protopopescu X, Butler T, Pan H et al (2008) Hippocampal structural changes across the menstrual cycle. Hippocampus 18:985–988
55. Oatridge A, Holdcroft A, Saeed N et al (2002) Change in brain size during and after pregnancy: study in healthy women and women with preeclampsia. Am J Neuroradiol 23:19–26
56. Love G, Torrey N, McNamara IM et al (2005) Maternal experience produces long-lasting behavioral modifications in the rat. Behav Neurosci 119:1084–1096
57. Ptok U, Barkow K, Heun R (2002) Fertility and number of children in patients with Alzheimer's disease. Arch Womens Ment Health 5:83–86
58. Colucci M, Cammarata S, Assini A et al (2006) The number of pregnancies is a risk factor for Alzheimer's disease. Eur J Neurol 13:1374–1377

59. McLay RN, Maki PM, Lyketsos CG (2003) Nulliparity and late menopause are associated with decreased cognitive decline. J Neuropsychiatry Clin Neurosci 15:161–167
60. Pawluski JL, Vanderbyl BL, Ragan K, Galea LAM (2006) First reproductive experience persistently affects spatial reference and working memory in the mother and these effects are not due to pregnancy or 'mothering' alone. Behav Brain Res 175:157–165
61. Lambert KG, Berry AE, Griffins G et al (2005) Pup exposure differentially enhances foraging ability in primiparous and nulliparous rats. Physiol Behav 84:799–806
62. Tulchinsky D, Hobel JH, Yeager E, Marshall JR (1972) Plasma estrone, estradiol, progesterone, and 17-hydroxyprogesterone in human pregnancy 1. Normal pregnancy. Am J Obstet Gynecol 112:1095–1100
63. Smith R, Mesiano S, McGrath S (2002) Hormone trajectories leading to human birth. Regul Pept 108:159–164
64. Wood CE (2005) Estrogen/hypothalamus-pituitary-adrenal axis interactions in the fetus: the interplay between placenta and fetal brain. J Soc Gynecol Investig 12:67–76
65. Storment JM, Meyer M, Osol G (2000) Estrogen augments the vasodilatory effects of vascular endothelial growth factor in the uterine circulation of the rat. Am J Obstet Gynecol 183:449–453
66. Ahdieh HB, Mayer AD, Rosenblatt JS (1987) Effects of brain antiestrogen implants on maternal behavior and on postpartum estrus in pregnant rats. Neuroendocrinology 46:522–531
67. Numan M, Insel TR (2003) The neurobiology of parental behavior. Springer, New York, pp 8–41
68. Numan M, Rosenblatt JS, Komisaruk BR (1977) Medial preoptic area and onset of maternal behavior in the rat. J Comp Physiol Psychol 91:146–164
69. Fahrbach SE, Pfaff DW (1986) Effect of preoptic region implants of dilute estradiol on the maternal behavior of ovariectomized nulliparous rats. Horm Behav 20:354–363
70. Maestripieri D, Zehr JL (1998) Maternal responsiveness increases during pregnancy and after estrogen treatment in macaques. Horm Behav 34:223–230
71. Pryce CR, Abbott DH, Hodges JH, Martin RD (1988) Maternal behavior is related to prepartum urinary estradiol levels in red-bellied tamarin monkeys. Physiol Behav 44:717–726
72. Pryce CR, Döbeli M, Martin RD (1993) Effects of sex steroids on maternal motivation in the common marmoset (*Callithrix jacchus*): development and application of an operant system with maternal reinforcement. J Comp Psychol 107:99–115
73. Kendrick KM, Keverne EB, Hinton MR, Goode JA (1986) Cerebrospinal fluid levels and acetylcholinesterase, monoamines and oxytocin during labour, parturition, vaginocervical stimulation, lamb separation and suckling in sheep. Neuroendocrinology 44:149–156
74. Dawood MY, Ylikorkala O, Trivedi D, Fuchs F (1979) Oxytocin in maternal circulation and amniotic fluid during pregnancy. J Clin Endocrinol Metab 49:429–434
75. De Geest K, Thiery M, Piron-Possoyt G, Vanden Driessche R (1985) Plasma oxytocin in human pregnancy and parturition. J Perinat Med 13:3–13
76. Silber M, Larsson B, Uvnas-Moberg K (1991) Oxytocin, somatostatin, insulin and gastrin concentrations vis-a-vis late pregnancy, breastfeeding and oral contraceptives. Acta Obstet Gynecol Scand 70:283–289
77. Levine A, Zagoory-Sharon O, Feldman R, Weller A (2007) Oxytocin during pregnancy and early postpartum: individual patterns and maternal-fetal attachment. Peptides 28:1162–1169
78. Buhimschi CS (2004) Endocrinology of lactation. Obestet Gynecol Clin North Am 31:963–979
79. Insel TR, Hulihan TJ (1995) A gender-specific mechanism for pair-bonding: oxytocin and partner preference formation in monogamous voles. Behav Neurosci 109:782–789
80. Williams JR, Insel TR, Harbaugh CR, Carter CS (1994) Oxytocin centrally administered facilitates formation of a partner preference in female pairie voles (*Microtus ochrogaster*). J Neuroendocrinol 6:247–250
81. Cho MM, DeVries CA, Williams JR, Carter CS (1999) The effects of oxtytocin and vasopressin on partner preferences in male and female prairie voles (*Microtus ochrogaster*). Behav Neurosci 113:1071–1080

82. Witt DM, Carter CS, Walton D (1990) Central and peripheral effects of oxytocin administration in prairie voles (*Microtus ochrogaster*). Pharmacol Biochem Behav 37:63–69
83. Cushing BS, Carter CS (1998) Peripheral pulses of oxytocin facilitate partner preference and increase probability of mating in female prairie voles. Soc Behav Neuroendocrinol Abstr 30:148
84. Cushing BS, Carter CS (2000) Peripheral pulses of oxytocin increase partner preferences in female, but not male, prairie voles. Horm Behav 37:49–56
85. Cushing BS, Martin JO, Young LJ, Carter CS (2001) The effects of peptides on partner preference formation are predicted by habitat in prairie voles. Horm Behav 39:48–58
86. Razzoli M, Cushing BS, Carter CS (2003) Hormonal regulation of agonistic and affiliative behavior in female mongolian gerbils (*Meriones unguiculatus*). Horm Behav 43:549–553
87. Pedersen CA, Prange AJJ (1979) Induction of maternal behavior in virgin rats after intracerebroventricular administration of oxytocin. Proc Natl Acad Sci U S A 76:6661–6665
88. Fahrbach SE, Morrell JI, Pfaff DW (1985) Possible role for endogenous oxytocin in estrogen-facilitated maternal behavior in rats. Neuroendocrinology 40:526–532
89. Insel TR, Harbaugh CR (1989) Lesions of the hypothalamic paraventricular nucleus disrupt the initiation of maternal behavior. Physiol Behav 45:1033–1041
90. Champagne F, Meaney MJ (2001) Like mother, like daughter: evidence for non-genomic transmission of parental behavior and stress responsitivity. Prog Brain Res 133:287–302
91. Pedersen CA, Boccia ML (2003) Oxytocin antagonism alters rat dams oral grooming and upright posturing over pups. Physiol Behav 80:233–241
92. Holman SD, Goy RW (1995) Experiential and hormonal correlates of caregiving in rhesus macaques. In: Pryce CR, Martin RD, Skuse D (eds) Motherhood in human and nonhuman primates: biosocial determinants. 3rd Schultz-Biegert Symposium, Kartause Ittingen, Switzerland, September 1994. Karger, Basel, Swizterland, pp 87–93
93. Seltzer LJ, Ziegler TE (2007) Non-invasive measurement of small peptides in the common marmoset (*Callithrix jacchus*): a radiolabeled clearance study and endogenous excretion under varying social conditions. Horm Behav 51:436–442
94. Maestripieri D, Hoffman CL, Anderson GM, Carter CS, Higley JD (2009) Mother-infant interactions in free-ranging rhesus macaques: relationships between physiological and behavioral variables. Physiol Behav 96:613–619
95. Carmichel MS, Humbert R, Dixen J, Palmisano G, Greenleaf W, Davidson JM (1987) Plasma OT increases in the human sexual response. J Clin Endocrinol Metab 64:27–31
96. Light KC, Grewen KM, Amico JA, Boccia ML, Brownley KA, Johns JM (2004) Deficits in plasma oxytocin responses and increased negative affect, stress, and blood pressure in mothers with cocaine exposure during pregnancy. Addict Behav 29:1541–1564
97. Light KC, Smith TE, Johns JM, Brownley KA, Hofheimer JA (2000) Oxytocin responsivity in mothers of infants: a preliminary study of relationships with blood pressure during laboratory stress and normal ambulatory activity. Health Psychol 19:560–567
98. Tops M, Van Peer JM, Korf J, Wijers AA, Tucker DM (2007) Anxiety, cortisol and attachment predict plasma oxytocin. Psychophysiology 44:444–449
99. Turner RA, Altemus M, Enos T, Cooper B, McGuinness T (1999) Preliminary research on plasma oxytocin in normal cycling women: investigating emotion and interpersonal distress. Psychiatry 62:97–113
100. Zak PJ, Kurzban R, Matzner WT (2005) Oxytocin is associated with human trustworthiness. Horm Behav 48:522–527
101. Heinrichs M, Baumgartner T, Kirschbaum C, Ehlert U (2003) Social support and oxytocin interact to suppress cortisol and subjective responses to psychosocial stress. Biol Psychiatry 54:1389–1398
102. Taylor SE, Klein LC, Lewis BP, Gruenwald TL, Gurung RA, Updegraff JA (2000) Biobehavioral responses to stress in females: tend-and-befriend, not fight-or-flight. Psychol Rev 107:411–429
103. Zak PJ, Stanton AA, Ahmadi S (2007) Oxytocin increases generosity in humans. PLoS One 7:e1128

104. Rimmele U, Hediger K, Heinrichs M, Klaver P (2009) Oxytocin makes a face in memory familiar. J Neurosci 7:38–42
105. Guastella AJ, Carson DS, Dadds MR, Mitchell PB, Cox RE (2009) Does oxytocin influence the early detection of angry and happy faces? Psychoneuroendocrinology 34:220–225
106. Guastella AJ, Mitchell PB, Dadds MR (2008) Oxytocin increases gaze to the eye region of human faces. Biol Psychiatry 63:3–5
107. Gordon I, Zagoory-Sharon O, Schneiderman I, Leckman JF, Weller A, Feldman R (2008) Oytocin and cortisol in romantically unattached young adults: associations with bonding and psychological distress. Psychophysiology 45:349–352
108. Ditzen B, Schaer M, Gabriel B, Bodenmann G, Ehlert U, Heinrichs M (2009) Intranasal oxytocin increases positive communication and reduces cortisol levels during couple conflict. Biol Psychiatry 65:728–731
109. Domes G, Heinrichs M, Michel A, Berger C, Herpertz SC (2007) Oxytocin improves "mind-reading" in humans. Biol Psychiatry 61:731–733
110. Rose JC, Schwartz J, Green J, Kerr DR (1998) Development of the corticotropin-releasing factor adrenocorticotropic hormone/beta-endorphin system in the mammalian fetus. In: Polin RA, Fox WW (eds) Fetal and neonatal physiology. W. B. Saunders, Philadelphia, pp 2431–2442
111. Winter SDW (1998) Fetal and neonatal adrenocortical physiology. In: Polin RA, Fox WW (eds) Fetal and neonatal physiology. W. B. Saunders, Philadelphia, pp 378–403
112. Mesiano S, Jaffe RB (1997) Developmental and functional biology of the primate fetal adrenal cortex. Endocr Rev 18(18):378–403
113. Heinrichs M, Meinlschmidt G, Neumann I et al (2001) Effects of suckling on hypothalamic-pituitary-adrenal axis responses to psychosocial stress in postpartum lactating women. J Clin Endocrinol Metab 86(10):4798–4804
114. Heinrichs M, Neumann I, Ehlert U (2002) Lactation and stress: protective effects of breast-feeding in humans. Stress 5:195–203
115. Douglas AJ, Brunton PJ, Bosch O, Russell JA, Neumann ID (2003) Neuroendocrine responses to stress in mice: hyporesponsiveness in pregnancy and parturition. Endocrinology 144:5268–5276
116. Neumann ID (2003) Brain mechanisms underlying emotional alterations in the peripartum period in rats. Depress Anxiety 17:111–121
117. Adam EK, Gunnar MR (2001) Relationship functioning and home and work demands predict individual differences in diurnal cortisol patterns in women. Psychoneuroendocrinology 26:189–208
118. Tu MT, Lupien SJ, Walker CD (2006) Multiparity reveals the blunting effect of breastfeeding on physiological reactivity to psychological stress. J Neuroendocrinol 18:494–503
119. Tu MT, Lupien SJ, Walker C-D (2006) Diurnal salivary cortisol levels in postpartum mothers as a function of infant feeding choice and parity. Psychoneuroendocrinology 31:812–824
120. Lankarani-Fard A, Kritz-Silverstein D, Barrett-Conner E, Goodman-Gruen D (2001) Cumulative duration of breast-feeding influences cortisol levels in postmenapausal women. J Womens Health Gend Based Med 10:681–687
121. DeVries AC, DeVries MB, Taymans S, Carter CS (1995) Modulation of pair bonding in female prairie voles (Microtus ochrogaster) by corticosterone. Proc Natl Acad Sci U S A 92:7744–7748
122. Storey AE, Walsh CJ, Quinton R, Wynne-Edwards KE (2000) Hormonal correlates fo paternal responsiveness in new and expectant fathers. Evol Hum Behav 21:79–95
123. Reburn CJ, Wynne-Edwards KE (1999) Hormonal changes in males of a naturally biparental and a uniparental mammal. Horm Behav 35:163–176
124. Hennessy MB, Harney KS, Smotherman WP, Coyle S, Levine S (1977) Adrenalectomy-induced deficits in maternal retrieval in the rat. Horm Behav 9:222–227
125. Rees SL, Panesar S, Steinger M, Fleming AS (2004) The effects of adrenalectomy and corticosterone replacement on maternal behavior in the postpartum rat. Horm Behav 46:411–419

126. Bardi M, French JA, Ramirez SM, Brent L (2004) The role of the endocrine system in baboon maternal behavior. Biol Psychiatry 55:724–732
127. Fleming AS, Steiner M, Anderson V (1987) Hormonal and attitudinal correlates of maternal behavior during the early postpartum period. J Reprod Infant Psychol 5:193–205
128. Stallings J, Fleming AS, Corter C, Worthman C, Steiner M (2001) The effects of infant cries and odors on sympathy, cortisol, and autonomic responses in new mothers and nonpostpartum women. Parent Sci Pract 1:71–100
129. Vale W, Spiess J, Rivier C, Rivier J (1981) Characterization of a 41-residue ovine hypothalamic peptide that stimulates secretion of corticotropin and β-endorphin. Science 213:1394–1397
130. Lowry PJ (1993) Corticotropin-releasing factor and its binding protein in human plasma. Ciba Found Symp 172:108–115
131. King BR, Smith R, Nicholson RC (2001) The regulation of human corticotrophin-releasing hormone gene expression in the placenta. Peptides 22:1941–1947
132. Scatena CD, Adler S (1998) Characterization of a human-specific regulator of placental corticotropin-releasing hormone. Mol Endocrinol 12:1228–1240
133. Raadsheer FC, Hoogendijk WJ, Stam FC, Tilders FJ, Swaab DF (1994) Increased numbers of corticotropin-releasing hormone expressing neurons in the hypothalamic paraventribular nucleus of depressed patients. Neuroendocrinology 60:436–444
134. Raadsheer FC, van Heerikhuize JJ, Lucassen PJ, Hoogendijk WJ, Tilders FJ, Swaab DF (1995) Corticotropin-releasing hormone mRNA levels in the paraventricular nucleus of patients with Alzhiemer's disease and depression. Am J Psychiatry 152:1372–1376
135. Yim IS, Gynn LM, Dunkel Schetter C, Hobel CJ, Chicz-DeMet A, Sandman CA (2009) Elevated corticotropin-releasing hormone in human pregnancy increases the risk of postpartum depressive symptoms. Arch Gen Psychiatry 66:162–169
136. Vanston CM, Watson NV (2005) Selective and persistent effect of foetal sex on cognition in pregnancy. NeuroReport 16:779–782
137. Lei ZM, Rao CV (2001) Neural actions of luteinizing hormone and human chorionic gonadotropin. Semin Reprod Med 19:103–109
138. Yaron Y, Lehavi O, Orr-Urtreger A et al (2002) Maternal serum HCG is higher in the presence of a female fetus as early as week 3 post-fertilization. Hum Reprod 17:485–489
139. Obiekwe BC, Chard T (1982) Human chorionic gonadotropin levels in maternal blood in late pregnancy: relationship to birthweight, sex and condition of the infant at birth. Br J Obestet Gynaecol 89:543–546
140. Santolaya-Forgas J, Meyer MJ, Burton BK, Scommegna A (1997) Altered newborn gender distribution in patients with low mid-trimester maternal serum human chorionic gonadotropin (MShCG). J Matern-Fetal Med 6:111–114
141. DiPietro JA, Irizarry RA, Costigan KA, Gurewitsch ED (2004) The psychophysiology of the maternal-fetal relationship. Psychophysiology 41:510–520
142. Johnson TRB, Jordan ET, Paine LL (1990) Doppler recordings of fetal movement: II. Comparison with maternal perception. Obstet Gynecol 76:42–43
143. Kisilevsky BS, Killen H, Muir DW, Low JA (1991) Maternal and ultrasound measurements of elicited fetal movements: a methodologic consideration. Obstet Gynecol 77:889–892
144. Herzenberg LA, Bianchi DW, Schröder J, Cann HM, Iverson GM (1979) Fetal cells in the blood of pregnant women: detection and enrichment by flouresence-activated cell sorting. Proc Natl Acad Sci U S A 76:1453–1455
145. Bianchi DW, Zickwolf GK, Weil GJ, Sylvester S, DeMaria MA (1996) Male fetal progenitor cells persist in maternal blood for as long as 27 years postpartum. Proc Natl Acad Sci U S A 93:705–708
146. Khosrotehrani K, Johnson KL, Cha DH, Salomon RN, Bianchi DW (2004) Transfer of fetal cells with multilineage potential to maternal tissue. J Am Med Assoc 292:75–80
147. Johnson KL, Nelson JL, Furst DE et al (2001) Fetal cell microchimerism in tissue from multiple sites in women with systemic sclerosis. Arthritis Rheum 46:1848–1854

148. Dawe GS, Tan X-W, Xiao Z-C (2007) Cell migration from baby to mother. Cell Adh Migr 1:19–27
149. Tan X-W, Liao H, Sun L, Okabe M, Xiao Z-C, Dawe GS (2005) Fetal microchimerism in the maternal mouse brain: a novel population of fetal progenitor or stem cells able to cross the blood-brain barrier? Stem Cells 23:1443–1452
150. Carleton A, Rochefort C, Morant-Oria J et al (2002) Making scents of olfactory neurogenesis. J Physiol Paris 96:115–122
151. Barker DJP (1998) Mothers, babies and health in later life. Harcourt Brace, Edinburgh
152. Richards M, Hardy R, Kuh D, Wadsworth MEJ (2001) Birth weight and cognitive function in the British 1946 birth cohort: longitudinal population based study. Br Med J 322:199–203
153. Barker DJP, Gluckman PD, Godfrey KM, Harding JE, Owens JA, Robinson JS (1993) Fetal nutrition and cardiovascular disease. Lancet 341:938–941
154. McCormack VA, Dos Santos Silva I, De Stavola BL, Mohsen R, Leon DA, Lthell HO (2003) Fetal growth and subsequent risk of breast cancer: results from long term follow up of Swedish cohort. Br Med J 326:248–251
155. Roseboom TJ, van der Meulen JH, Ravelli AC, Osmond C, Barker DJ, Bleker OP (2001) Effects of prenatal exposure to the Dutch famine on adult disease in later life: an overview. Mol Cell Endocrinol 185:93–98
156. Beydoun H, Saftlas AF (2008) Physical and mental health outcomes of prenatal maternal stress in human and animal studies: a review of the recent evidence. Paediatr Perinat Epidemiol 22:438–466
157. Glynn LM, Sandman CA (2006) The influence of prenatal stress and adverse birth outcome on human cognitive and neurological development. In: Glidden L (ed) International review of research in mental retardation, vol 32. Academic, San Diego, CA, pp 110–122
158. O'Donnell K, O'Conner TG, Glover V (2009) Prenatal stress and neurodevelopment of the child: focus on the HPA axis and role of the placenta. Dev Neurosci 31:285–292
159. Bell RQ (1968) A reinterpretation of the direction of effects in studies of socialization. Psychol Rev 75:81–95
160. Boer GJ (1993) Chronic oxytocin treatment during late gestation and lactation impairs development of rat offspring. Neurotoxicol Teratol 15:383–389
161. Davis EP, Glynn LM, Dunkel Schetter C, Hobel C, Sandman CA (2005) Maternal plasma corticotropin-releasing hormone levels during pregnancy are associated with infant temperament. Dev Neurosci 27:299–305
162. Davis EP, Glynn LM, Schetter CD, Hobel C, Chicz-DeMet A, Sandman CA (2007) Prenatal exposure to maternal depression and cortisol influences infant temperament. J Am Acad Child Adolesc Psychiatry 46:737–746
163. Ellman LM, Dunkel Schetter C, Hobel CJ, Glynn LM, Sandman CA (2008) Timing of fetal exposure to stress hormones: effects on newborn physical and neuromuscular maturation. Dev Psychobiol 50:232–241
164. Bakker J, De Mees C, Douhard Q et al (2006) Alpha-fetoprotein protects the developing fetal mouse brain from masculinization and defeminination by estrogens. Nat Neurosci 9:220–226

Chapter 5
Maternal Thyroid Function During Pregnancy: Effects on the Developing Fetal Brain

Joanne F. Rovet and Karen A. Willoughby

Keywords Thyroid hormone • Thyroid development • Maternal hypothyroidism • Hypothyroxinemia • Pregnancy • Prenatal brain development • Cognitive development

Introduction

During pregnancy, numerous hormonal changes and increased metabolic demands lead to complex changes in maternal thyroid physiology and fetal health. Findings from both animal and human research convincingly show that thyroid hormone (TH) is essential for normal brain development and is important for the regulation of a number of critical neurobiological processes [1, 2]. This research has furthermore demonstrated that permanent neuropsychological deficits and alterations in brain development can occur if TH levels are insufficient during gestation [3–5].

Remarkably, it is not until the third trimester that the fetus produces its own TH in appreciable amounts and not until term that full thyroid function is achieved [6–8]. However, the fetal brain requires TH throughout gestation, including the period before the onset of fetal thyroid function [9, 10]. Consequently, the fetus must rely entirely on the mother's supply of TH during the first trimester [11–13]. In fact, autopsy studies have reported that measurable amounts of TH of maternal origin have been found in fetal brain tissue as early as the fifth week of gestation [8, 14, 15]. Even though the fetal thyroid undergoes substantial development through the course of gestation and produces increasingly larger quantities of its own TH, maternal supplementation continues to occur in order to offset any TH

J.F. Rovet (✉)
Hospital for Sick Children, University of Toronto, 555 University Avenue, Toronto, ON, M5G1X8, Canada
e-mail: joanne.rovet@sickkids.ca

A.W. Zimmerman and S.L. Connors (eds.), *Maternal Influences on Fetal Neurodevelopment: Clinical and Research Aspects*,
DOI 10.1007/978-1-60327-921-5_5, © Springer Science+Business Media, LLC 2010

insufficiencies associated with an immature fetal thyroid system [7, 16, 17]. Thus, for normal brain development and neurodevelopment to occur, an adequate supply of maternal TH is necessary throughout gestation, especially early in pregnancy [7].

In a substantial proportion of pregnancies, however, maternal TH levels are insufficient owing to some form of maternal thyroid dysfunction [18–20]. Recently, several studies have demonstrated that maternal TH insufficiency contributes to a number of adverse outcomes, including an increased incidence of reproductive problems and obstetrical complications [21–23], as well as suboptimal child neurodevelopment [4, 5]. Thus, the purpose of this chapter is to review current evidence of the action of TH in development and the effects on the progeny during gestation, and after birth if the maternal thyroid supply is inadequate.

Basic Thyroid Physiology

Thyroid hormone refers to two basic iodine-containing compounds, triiodothyronine (T3) and tetraiodothyronine or thyroxine (T4), which contain three and four iodide molecules, respectively [2, 24]. T3 and T4 are manufactured within the follicular cells of the thyroid gland, a double-lobed organ located in the base of the neck on either side of the trachea [24, 25]. The synthesis of these two hormones is dependent on an adequate supply of exogenously derived iodine [26] and involves a two-step process (for review see Bernal and Nunez [1]). The first step is the iodination of thyroglobulin, a large glycoprotein stored in the thyroid gland, to form mono- and di-iodothyronine. The second step involves the coupling of these two residues to form T3 or T4. T4 is usually the more abundant of the two hormones with approximately ten times more T4 than T3, which is being secreted by the thyroid into the bloodstream [24, 27]. In the brain, however, T3 is the main bioactive form of TH [3, 24].

Synthesis of T3 and T4 by the thyroid is regulated by a negative feedback system involving the hypothalamus and pituitary gland, and this system is known as the hypothalamic–pituitary–thyroid axis [1, 24, 25]. The hypothalamus first responds to the environmental cues by synthesizing and releasing thyrotropin-releasing hormone (TRH), which then binds to the receptors in the anterior pituitary gland and causes the synthesis of thyrotropin or thyroid stimulating hormone (TSH). TSH, which is released from the pituitary, then enters the bloodstream and travels to the thyroid gland, where it binds to specific receptors on thyroid cell membranes and stimulates the release of T3 and T4 from the thyroid gland [25, 28]. Once in the bloodstream, these two hormones travel via binding proteins (thyroid-binding globulin and albumin) to various target sites in the body, including the brain [24]. Most TH circulating in the bloodstream is bound to thyroid-binding proteins and thus is biologically inactive; however, a small percentage of circulating TH, known as free T4 (fT4) or free T3 (fT3), becomes unbound from the binding proteins and is available to elicit thyroid action [24]. If T4 or T3 levels are low, as in hypothyroidism, the HPT-axis will activate and stimulate the production and release of TRH and TSH and, in turn, increase the production of T3 and T4 [24, 29]. Alternatively, when circulating TH levels are high, the hypothalamus and pituitary produce less TRH and TSH, leading to an inhibition of T3 and T4 production by the thyroid [24]. Thus, this complex and tightly regulated feedback mechanism serves to ensure adequate synthesis of T3 during early brain development.

TH action is also controlled locally by three deiodinase (D) enzymes, D1, D2, and D3 [30–32]. These enzymes, which are produced in the liver, act locally and in a region- and time-specific manner, with only D2 and D3 being found in the brain [33, 34]. D2 acts by extracting one iodide molecule from the outer ring of the T4 molecule to produce T3, thereby facilitating production of this hormone. D3, in contrast, is found in placenta and brain and serves to inactivate T4 and its product T3, by deiodination of iodothyronines at the tyrosyl ring, thus forming *reverse* T3 (an isomer of T3) and T2, respectively [30, 33]. If T4 levels are low, D2 activity becomes upregulated and D3 activity is downregulated [30, 35]. This action serves as a protective mechanism during states of TH deficiency [30, 35]. Within the brain, D2 and D3 differ with respect to tissue distribution, developmental timing of expression (e.g., D3 is expressed earlier than D2), and the role each plays in increasing or decreasing T3 bioavailability [24, 25, 30, 33]. D2 is found mainly in astrocytes, whereas D3 is found predominantly in neuronal cells that contain thyroid receptors [2, 30] (Fig. 5.1).

T3 exerts its major actions within neurons directly by acting as a transcription factor, which regulates the expression of specific genes that are responsible for critical neurobiological events [3, 25, 30, 32]. This action is accomplished by the forma tion of a TH receptor complex, which is composed of other proteins, and together with T3 serves as a coactivator or corepressor for particular target genes [32].

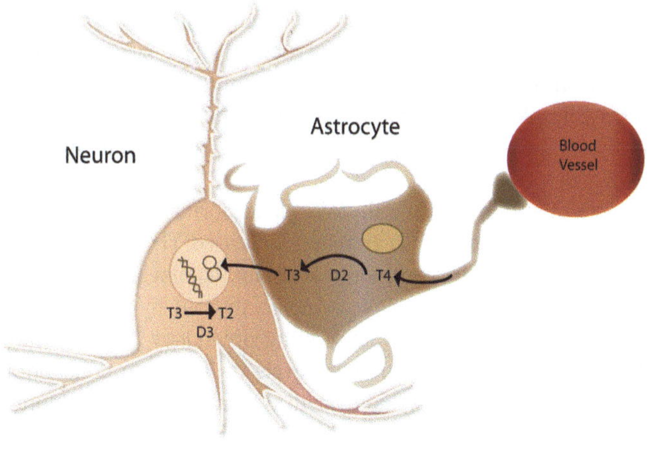

Fig. 5.1 Hypothesis for the transfer of T4 from blood vessels to astrocytes, where T4 is deiodinased by D2 to form T3. T3 is then transported to neurons that contain nuclear receptors, as well as D3, which inactivates T3 by forming T2. Reprinted from Santisteban and Bernal [36], with permission from Springer

More specifically, the TH receptor complex serves to up- or downregulate gene expression, thereby enabling production of critical proteins for brain development. Within the brain, both the exact time of action and specific site of action vary among different genes because they act in an orderly sequence to allow normal brain development [1, 25, 30, 32]. In conditions of low TH, dormant TH receptors can act as repressors of gene transcription, interfering with genetically determined developmental processes [36–38].

Thyroid Hormones in Pregnancy

Over the normal course of pregnancy, maternal TH concentrations undergo significant fluctuation due to increased requirements by the fetus for maternal T4. Typically, maternal T4 concentrations rise in early pregnancy and return to normal levels after delivery, thus ensuring that an adequate supply of maternal T4 reaches the fetus [11, 17, 38]. In the first trimester, increased maternal T4 production usually results in a corresponding decrease in circulating TSH levels because of altered homeostatic settings of the HPT axis [11, 39, 40]. Also, elevated estrogen levels in early pregnancy can lead to an increased amount of circulating TH binding proteins in the blood, enabling TH to travel more extensively [41]. Evidence suggests that these normal, but significant, fluctuations in TH concentrations in early pregnancy are biologically relevant for fetal neurodevelopment, as they maintain the required TH levels during critical periods of development [7, 42].

As mentioned previously, research on human embryos indicates that as early as 4 weeks after conception, circulating TH of maternal origin exists in embryonic fluids [8, 15]. This signifies that maternal TH is available via the placenta for fetal brain development from the second month of pregnancy [7]. The action of placental deiodinase enzymes within the placenta, particularly D3, serves to control the rate of TH transfer from mother to fetus, thus maintaining the requisite fetal TH levels [16, 43]. Following placental transfer, maternal T4 exerts its effects on the developing brain by binding to thyroid receptors and affecting gene expression in the fetal brain [44–46].

In healthy pregnant women, the thyroid gland normally maintains a functional reserve throughout pregnancy, thus enabling increased TH output and maintaining thyroid homeostasis [47]. However, some pregnant women are unable to meet these increased TH demands of pregnancy owing to thyroid hypofunction, which can either predate the pregnancy or develop during pregnancy [39, 48]. In such situations, the fetus is exposed to insufficient levels of TH from conception until the third trimester when the fetal thyroid system typically assumes a more prominent role in TH production [20].

Approximately 0.3–0.5% of pregnancies are characterized by overt hypothyroidism, defined as a higher than normal TSH level and a lower than normal T4 level, while in a further 2–3% of pregnancies, women experience subclinical hypothyroidism or only elevated TSH levels [49–51]. In addition, euthyroid pregnant women, whose fT4 levels are in the lowest tenth percentile and TSH levels are normal, may also be at risk for having offspring with impaired neurodevelopment because of early TH insufficiency [52]. Overt hypothyroidism in pregnancy has a wide range of causes including congenital or acquired hypothyroidism of youth, Hashimotos thyroiditis, thyroid gland ablation following Graves' disease or thyroid cancer, or damage to the pituitary gland by a tumor, radiation, or surgery [53, 54]. Although most causes of hypothyroidism develop prior to the pregnancy, a small proportion of women also develop hypothyroidism during pregnancy, with the most common cause being autoimmune thyroid disease [53]. Since prevalence rates for the various hypothyroid conditions differ depending on criteria used to diagnose hypothyroidism during pregnancy [18], rates can vary from as low as 0.19% in Japan [55] to 2.2% in Belgium [39] and 2.5% in the United States [56] (Table 5.1).

Among nonpregnant patients, hypothyroidism is usually defined as a TSH level of 4–6 μU/L or higher and a fT4 value between 9 and 23 mIU/L [51, 57], depending upon the laboratory used. However, there are no standard normative values to diagnose hypothyroidism in pregnancy, and normal physiologic changes in thyroid function during pregnancy often make the diagnosis of maternal hypothyroidism difficult. Nevertheless, experts in this field do recommend using trimester-specific norms [54, 58–60]. Generally, population-specific and gestational age-corrected thresholds of TSH are thought to be more reliable and sensitive indices of thyroid deficiency than fT4 levels, and also more predictive of hypothyroidism up to 5 years later [4, 50, 61, 62].

Hypothyroidism is typically treated using synthetic TH (Levothyroxine or LT4 or Synthroid) to supplement the body's insufficient TH levels and mimic the normal changes in TSH and T4 concentrations observed in healthy pregnant women [47]. Since it can take several weeks from the onset of treatment to achieve adequate TH levels, it is strongly recommended that hypothyroid pregnant women be tested and medicated as early as possible in order to ensure that TH levels are stabilized early in pregnancy and fetal TH needs are continuously met [63]. According to Alexander et al. [48], about 85% of pregnant women, who are monitored for hypothyroidism before pregnancy, exhibit high TSH concentrations at 8–12 weeks of gestation and require an increase of approximately 50% in their LT4 dosage in the first half of pregnancy. Unfortunately, this increased dose is typically not given to most pregnant hypothyroid women, and therefore, a large majority of their offspring will experience a period of TH insufficiency in the first half of pregnancy [48]. Consequently, it is recommended that all women with overt hypothyroidism have their TH levels closely and frequently monitored and their treatments modified as needed [48, 54, 64].

Table 5.1 Definitions of thyroid dysfunction

Function	TSH	Free T_4
Clinical hypothyroidism	Increase	Decrease
Subclinical hypothyroidism	Increase	Normal
Hypothyroxinemia	Normal	Decrease

TSH thyroid stimulating hormone; T_4 thyroxine

Unless adequate treatment is provided, most women with maternal hypothyroidism will likely provide their fetuses with insufficient TH at some point during gestation.

Pregnant women with subclinical hypothyroidism who experience only TSH elevations rarely show signs or symptoms of thyroid dysfunction [51, 65–67] and therefore, treatment of subclinical hypothyroidism is seldom provided during pregnancy [50]. Although many of these women do show a spontaneous return to normal TSH levels after approximately 10 weeks of gestation, irreversible neurodevelopmental damage to the fetus can still occur during this early period [47].

Given that maternal iodine intake is essential for both maternal and fetal TH synthesis and that metabolism of iodine is substantially increased during pregnancy in order to meet fetal TH demands, maternal iodine deficiency can also lead to hypothyroidism during pregnancy [41, 68, 69]. In cases of severe iodine deficiency, synthesis of TH is inhibited leading to development of a goiter (i.e., swelling of the thyroid gland), as the mother's thyroid gland enlarges so as to trap more iodine. Iodine deficiency is most prevalent in areas where the soil has been depleted of iodine, particularly parts of Latin America, Asia, and Africa, thus making iodine deficiency the most common preventable cause of hypothyroidism worldwide [7, 41, 69]. Severe maternal hypothyroxinemia due to iodine deficiency during pregnancy can result in severe mental and physical retardation in the offspring, a condition known as neurological cretinism [2, 7, 20, 70]. Interestingly, it has been reported that in recent years, iodine levels have been declining in the United States due to the elimination of iodophors in the dairy and wheat industries, and this is a cause for concern [69]. In addition, the lack of iodine in most prenatal vitamins within the United States may also lead to insufficient iodine levels during pregnancy [70]. Overall, iodine supplementation in pregnancy is recommended to ensure adequate production of TH in the mother and fetus [56, 71].

Thyroid Hormones and Fetal Brain Development

Fetal thyroid gland development begins around the 20th day of gestation with the emergence of a small rudimentary gland that migrates from its initial position at the base of the tongue to its final location in the neck by gestational days 45–50 [24, 72]. By the tenth week of gestation, the thyroid gland begins concentrating iodide and producing thyroglobulin (Tg), the precursor protein from which TH is formed [24, 72–74]. Although the fetal thyroid acquires the capacity to synthesize TH by 10–12 weeks of gestation, significant fetal TH production does not occur until 20 weeks of gestation [17, 24, 75]. In addition, hypothalamo-pituitary regulation of fetal TH production does not occur until the third trimester and concentrations of TH do not approach adult levels until birth [17, 24, 74, 75]. Several groups have also reported that receptors for TH (particularly for T3) are expressed in the fetal brain before the onset of fetal thyroid function and increase rapidly from 10 to 16 weeks of gestation during the period of active cortical neurogenesis [44, 45, 76].

A number of factors determine the concentration levels of T4 in the developing brain and placenta, including dietary iodine and effective regulation of the activity of the maternal thyroid gland by the hypothalamus [1]. Importantly, the presence of TH in fetal brain tissue as early as 9–12 weeks of gestation [14] supports the notion that most T4 in fetal tissues in early gestation is maternal in origin, and maternal TH continues to play a critical role throughout gestation until transfer of maternal TH discontinues at birth [7, 15, 76, 77].

To elucidate the role of thyroid hormones in fetal brain development and function and to gain a more extensive understanding of the timing of TH actions, in vitro and in vivo animal models of hypothyroidism are typically used (for reviews see [47] and [13]). For example, studies using cultured rat cortical neurons have shown that T3 promotes synapse formation in the cerebral cortex [78]. Numerous animal studies, along with clinical evidence of the effects of congenital hypothyroidism (i.e., fetal thyroid dysfunction) and iodine deficiency, indicate that TH is essential in regulating neurodevelopment and its actions follow a strict region and time-specific pattern [1, 9, 25, 30, 47, 79, 80] (Fig. 5.2).

In the developing nervous system, the predominant effect of TH is at a cellular level, where several fundamental neurodevelopmental processes, such as neurogenesis, neuronal proliferation and migration, and axonal and dendritic growth, are regulated [10, 42, 81–84]. Animal studies show that TH also plays a significant role in synaptogenesis and myelination during brain development, as well as modulation of hippocampal neurogenesis in the adult brain [3, 42, 85, 86]. Moreover, TH is involved in the development of different neurotransmitter systems and is necessary for subsequent neurotransmitter functioning [87, 88]. In fact, evidence suggests that there may be a link between low TH levels and several disorders, including psychiatric illnesses (mood disorders and depression) and neurodegenerative diseases (such as frontotemporal dementia and Alzheimer's disease) [89–91].

Within the brain, TH acts by modulating the changes in gene products via alterations in the transcription of target genes, which are active during different critical periods of human brain development and are expressed in different brain structures [1, 92]. For instance, TH-regulated genes are found within the hippocampus, hypothalamus, pituitary gland, cerebral cortex, corpus callosum, and cerebellum [1, 2, 88]. In addition, Kester et al. [30] have reported an increase in T3 levels and activity of the D2 enzyme in several different brain regions during gestation, including the fetal cerebral cortex. Evidence also indicates that the actions of TH in these brain regions follow a strict temporal and regionally specific schedule, with most TH-dependent genes being sensitive to TH only during a finite and very limited time period [32, 92].

Given the many effects of TH on the developing nervous system, TH deficiency can have a significant impact on neurodevelopmental processes leading to permanent alterations in the anatomy and function of the central nervous system [1, 66, 81, 93, 94]. For example, animal research has shown that maternal hypothyroidism causes delays in the neurodevelopment of neonatal rats [95], and alters the expression of specific genes such as those coding for neuroendocrine-specific protein (NSP) and Oct-1, proteins involved in cell proliferation in the rat brain [42, 96].

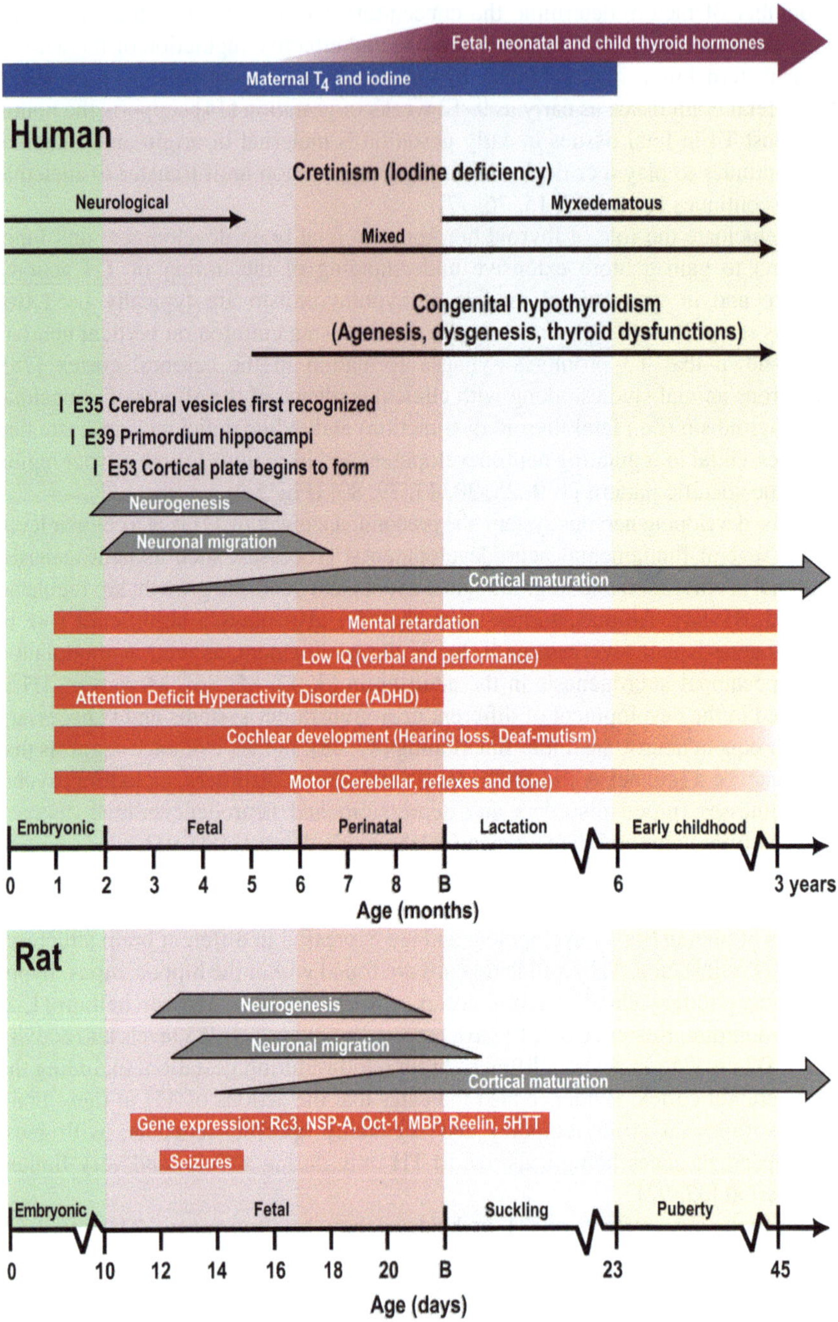

Fig. 5.2 Overall neurodevelopmental events and neurological alterations associated with iodine (and thyroid) deficiency during fetal pre- and postnatal life. The period in which both T4 and iodine are transferred from the mother to the fetus, and the period in which the fetal thyroid gland begins

In addition, hypothyroidism is associated with a reduction in myelination, which at a later age can have a significant impact on neuronal connectivity and establishment of neuronal networks [93, 97]. Of note, Goodman and Gilbert [83] recently found abnormal cortical development in rats even after only modest decreases in maternal T4 during the prenatal period.

It is generally agreed that the timing and duration of TH deficiency in pregnancy has a significant impact on the type and severity of neurological disruption [80, 98]. Given knowledge of critical periods in early and late gestation during which distinct parts of the brain are sensitive to TH [7], a deficient supply of TH during these critical periods can result in the delayed onset of important biological events and can cause irreparable harm to the developing brain [8, 30, 80]. In humans, two main waves of cell migration occur in the neocortex before midgestation, with peaks at 8–10 weeks and 12 weeks of gestation, respectively [99]. Although few studies have investigated the neurodevelopmental effects of TH deficiency during early gestation in humans, recent animal studies have shown that mild reductions in maternal TH in pregnant rats during early gestation disrupts early migrational direction of neurons in the fetal cortex and hippocampus and leads to aberrant location of neurons in adult offspring brain [42, 81, 100].

Maternal hypothyroidism is also known to interfere with several neurodevelopment processes occurring before gestational day 20 in the rat, including neuronal proliferation [101] and migration of cells [102]. In the cerebral cortex, TH deficiency results in impaired cortical layering and altered callosal connections [102, 103]. White matter tracts, predominantly populated by neuronal axons and myelinating oligodendrocytes, are also significantly reduced in size in the hypothyroid rat brain [104]. Sharlin et al. [105] have shown that maternal hypothyroidism can also alter the balance between astrocyte and oligodendrocyte formation, favoring the formation of astrocytes and leading to reduced myelin production. In fact, in the rat, the size of corpus callosum and anterior commissure, the brain's two largest white matter tracts, are both substantially reduced following maternal hypothyroidism [81, 103, 104].

The hippocampus, which is an essential structure for memory [106] and visuospatial learning [107], is particularly vulnerable to TH insufficiency [108, 109]. For instance, several studies indicate that both the number of granule cells in the dentate gyrus and pyramidal cells in CA1 of the hippocampus are irreversibly reduced in TH-deficient rats [87, 108, 110]. TH-deficiency is also associated with

Fig. 5.2 (continued) to produce its own TH are indicated at the *top* of the figure. Timing of developmental periods and major developmental events of the cerebral cortex are indicated for both humans (*top*) and rats (*bottom*). Thyroid hormone deficiencies and their etiologies are shown in the upper part of the human panel, and *thin black arrows* represent the crucial periods related to these disorders. Finally, *rectangle bars* in the human panel indicate some neurological alterations and vulnerability periods associated with maternal/fetal T4 and iodine deficiency, and *rectangle bars* in the lower rat panel indicate genes that are regulated by TH and behavioral alterations associated with maternal and/or fetal T4 and iodine deficiency. Reprinted from Berbel et al. [102], with permission from Elsevier

disrupted synaptic functioning in the CA1 region and dentate gyrus of the hippocampus [87, 88, 111, 112]. Furthermore, animal research indicates a dose-dependent effect of TH-deficiency on hippocampal development, resulting in greater abnormalities in both hippocampal structure and spatial learning and memory abilities in rats with severe TH deficiencies vs. rats with low-to-moderate TH deficiencies [113].

Within the cerebellum, TH is known to directly regulate several critical developmental genes, such as those coding for neurotrophins and Purkinje cell protein-2, involved in Purkinje cell formation [2, 3, 114]. Purkinje cells, which are inhibitory neurons that play a vital role in cerebellar functioning by relaying information from the cerebellum to various other brain regions, exhibit a decreased rate and delayed period of cell differentiation following neonatal hypothyroidism. Delayed proliferation and migration of cerebellar granule cells is also observed in TH-deficient rats [3, 115], as is delayed myelination of the cerebellum and reduced synaptogenesis between Purkinje cells and granule cell axons [2, 116]. Finally, several studies examining the effects of TH deficiency in other regions of the rat brain have reported impaired axonal maturation within the corpus callosum [102, 117] and abnormal development of pyramidal cells within the visual cortex [118, 119].

In summary, TH appears to have a significant impact on the cytoarchitecture of many brain regions through its effects on cell proliferation and migration, synaptogenesis and myelination. Thus, TH is critical for brain development and insufficient TH levels during pregnancy may result in the abnormal development of brain structures that are essential for certain cognitive abilities later in life.

A Historical Perspective on Maternal Hypothyroidism

An awareness of the association between the thyroid gland and brain development has been evident since the middle of the nineteenth century following the first reports of neurological cretinism [120]. Cretinism, a syndrome associated with mild to severe mental retardation, stunted physical growth, and abnormal brain development, results from untreated pre- and postnatal TH-insufficiency and prolonged nutritional iodine deficiency [93]. While cretinism has been almost completely eliminated in the developed world due to greater dietary intake of iodine, in the eighteenth and nineteenth centuries, cretinism was often associated with development of a goiter (substantial thyroid enlargement) [121]. Despite speculations that an endemic goiter signified thyroid gland degeneration, many physiologists in the late nineteenth century had a poor understanding of thyroid function as well as the causes of goiters and cretinism [121]. For instance, Flint [122] reported in the third volume of *The Physiology of Man*:

> It is generally admitted that the thyroid gland may be removed from animals without interfering with any of the vital functions; and this taken in connection with the fact that it is so often diseased in the human subject, without producing any general disturbance, shows that its function cannot be very important. Nothing of importance has been learned from a chemical analysis of its substance.

However, by the early twentieth century, physicians familiar with endemic goiter and cretinism began to recognize that the developing fetal thyroid may be "susceptible to influences which impair the mother's thyroidal resources" [123]. In fact, Hunziker-Shild [124] hypothesized that the mother may be the only source of TH, given that children born without a thyroid gland (i.e., athyrotic) only begin to show symptoms of cretinism a few months after birth [20]. Early research on thyroid development in humans was based largely on histological studies of miscarried or aborted fetuses and examination of blood samples obtained in early pregnancy or at delivery [7, 125, 126].

Research on thyroid function improved significantly with the identification of thyroxine (T4) by Harington and Barger [127] in 1927 and triiodothyronine (T3) by Gross and Pitt-Rivers [128] in 1952, as well as the discovery that low circulating maternal T4 levels were related to neurological cretinism in the offspring [129, 130]. Eventually these advances, as well as major technical improvements in detecting TH levels, led to the implementation of neonatal TH screening and treatment programs, which effectively prevent the severe brain damage associated with congenital hypothyroidism. Unfortunately, the success of these early TH treatment programs led many researchers to believe that the effects of TH on brain development occurred only after birth, because minimal damage to the central nervous system was observed if athyrotic newborns were treated promptly with T4 [7, 25, 79]. In addition, due to early evidence indicating that the placenta was impermeable to TSH, some researchers believed that the placenta also prevented significant transfer of maternal T4 to the fetus during gestation [20, 131].

In the late 1960s and early 1970s, Man and colleagues [132], while conducting the Rhode Island Lying-In Study of late maternal age, were able to measure serum butanol-extractable iodine levels (BEI; the only measure of circulating TH available at the time) in pregnant women participating in the study. Later on, the offspring of these participants were assessed in infancy and at 4 and 7 years of age on various measures. Comparisons between children whose mothers had low BEI levels during pregnancy and those whose mothers had normal levels indicated that the former group had (1) an increased incidence of subnormal intelligence, (2) disabilities in visuospatial and locomotor domains, (3) increased inactivity, and (4) slow reaction times [132]. Unfortunately, these results were not fully appreciated at the time, given prevailing views that early neurodevelopment depended solely on the fetal supply of TH and that maternal TH did not cross the placenta (see Chan and Rovet [47]).

In the late 1980s, however, Vulsma and colleagues [12] produced the first clear evidence that maternal T4 does cross the placenta in late gestation. These investigators observed moderate levels of T4 (i.e., 30–60% of normal values) in cord sera of neonates who were not able to synthesize any T4 on their own, because of thyroid agenesis or a total organification defect [12]. Also observed were significant declines in their T4 levels in the postpartum period [12]. Additionally, in the late 1980s and early 1990s, several studies showed evidence of maternal T4 in coelomic and amniotic fluids as early as the fifth week of gestation, and that these concentrations (a) increased steadily during the first half of pregnancy and (b) correlated significantly with circulating maternal levels of T4 [8, 15].

Studies on Outcome Following Maternal Hypothyroidism

To date, a large number of studies have shown that disturbances in maternal thyroid function are associated with multiple adverse pregnancy outcomes, including preeclampsia, placental disruptions, pregnancy-induced hypertension, fetal distress in labor, caesarean section, fetal death, and postpartum hemorrhage [39, 51, 63, 64, 66, 133, 134]. Early miscarriage, preterm delivery, and lower birth weight in offspring are also common in hypothyroid women [18, 21, 23, 51, 54]. Typically, women with subclinical hypothyroidism or adequately treated clinical hypothyroidism have fewer complications during pregnancy than those with inadequately treated overt hypothyroidism [57, 63]. Thus, early diagnosis and treatment of maternal hypothyroidism is critical for both the mother and her offspring. However, it should be noted that in a study, Casey et al. [135] found no evidence of improved pregnancy outcome following treatment for either maternal hypothyroxinemia or subclinical hypothyroidism during pregnancy, whereas Negro et al. [136] found that LT4 therapy lowered the chances of miscarriage and premature delivery in pregnant women positive for thyroid peroxidase antibodies and who later developed thyroid dysfunction. Finally, Dussault and Fisher [137] found that subclinical maternal hypothyroidism was associated with an increased rate of congenital hypothyroidism in offspring, possibly because of disruption of the fetal thyroid gland by the transfer of maternal TSH receptor-blocking antibodies.

Over the past 10 years, considerable interest has been generated in maternal hypothyroidism, especially in light of two highly publicized studies from the United States and the Netherlands. These studies linked maternal thyroid hypofunction to neuropsychological [4] and psychomotor [5] impairment in the offspring. The Haddow et al. [4] study from Maine compared 7- to 9-year-old children whose mothers had elevated TSH levels (as determined from stored serum samples originally derived for alphafetoprotein testing) at 16 weeks of pregnancy with children of women with normal TSH levels during pregnancy on a range of neuropsychological tests. Children of the women with subclinical hypothyroidism attained IQ scores four points below controls, which was not statistically significant [4]. However, when only the children of women with high TSH values who did not receive treatment were compared with controls, IQ scores were shown to differ significantly by seven points, with over 19% of the untreated group scoring below 85 [4]. Furthermore, children of untreated hypothyroid women also attained lower scores on tests of language, attention, and learning, and had poor overall school performance when compared with controls. These findings, therefore, suggest that long-term neurocognitive deficits exist in the progeny of women not treated for subclinical hypothyroidism during pregnancy, especially during the first trimester [4].

Concurrently, a series of studies from the Netherlands by Pop and colleagues [5, 52] examined the effects of gestational hypothyroxinemia (i.e., low T4 levels and normal TSH levels) on the offspring. In their 1999 study, Pop et al. [5]. found that infants of women who experienced untreated hypothyroxinemia during the first trimester of pregnancy had significantly delayed motor and mental function, whereas hypothyroxinemia at 32 weeks of gestation had no effect. A second study examining child development at 1 and 2 years of age after maternal hypothryoxinemia found that neuropsychological development was most affected if maternal TH-deficiency was not corrected by 24 weeks of gestation [52]. This study provided important new evidence that early correction of maternal hypothyroxinemia can prevent the adverse child outcomes associated with TH insufficiency [52].

Several additional studies have reported suboptimal development in children exposed to hypothyroidism during pregnancy. Kooistra et al. [138] compared 108 neonates born to women with low maternal fT4 values with 96 neonates of women with normal fT4 values. At 3 weeks of age, the infants of hypothyroxinemic women scored lower in terms of orientation, and this ability was predicted by first trimester maternal fT4 values. Smit et al. [139] studied a small number of infants born to women with clinical hypothyroidism, who were grouped according to whether or not their mothers were overtly hypothyroid during the pregnancy. The children were assessed at 6, 12, and 24 months of age using the Bayley Scales of Infant Development. Those born to the women with overt hypothyroidism scored lower on the Bayley Mental Development Index than did those whose mothers were euthyroid during pregnancy [139]. In our laboratory, Mirabella and colleagues [140, 141] observed vision abnormalities, reflecting weaker contrast sensitivity, and decreased attention in offspring born to hypothyroid women vs. control women. Using a survey technique, Matsura and Konishi [142] reported that 80% of children born to women with severe hypothyroidism during pregnancy had developmental delays. In contrast Liu et al. [143] reported that the IQ scores of eight offspring of hypothyroid mothers who were treated adequately during the first half of pregnancy and early for their hypothyroidism did not differ from siblings born prior to the mother's hypothyroidism, thereby signifying no adverse effects of maternal hypothyroidism on offspring's mental development.

The timing and severity of TH insufficiency appears to predict type and severity of neurological deficits [9, 80, 85]. For instance, Man et al. [132] showed that children born to women with untreated maternal hypothyroxinemia before 24 weeks of gestation had visuospatial and motor deficits at ages 4 and 7, suggesting that the first 12–29 weeks of pregnancy may represent a critical period during which the neural substrates for abilities that depend on intact visual and motor systems, such as visual attention, visual processing, and motor skills, require TH [80, 132]. However, it should be noted that a recent study by Oken et al. [68] failed to find any meaningful associations between early maternal thyroid dysfunction in pregnancy and children's performance on tasks of visual recognition memory, visual motor ability, and receptive vocabulary. Interestingly, studies of later TH deficiency in pregnancy (e.g., congenital hypothyroidism) show that children with congenital hypothyroidism have an increased incidence of memory and learning deficits as well as subnormal visual (contrast sensitivity) and visuospatial abilities and fine motor skills [80]. Indeed, recent unpublished studies from our lab show that among offspring of women treated for overt hypothyroidism, those born to women who remained hypothyroid during pregnancy had lower scores on tests of visual ability, attention, and memory than did children born to women whose hypothyroidism was fully corrected during pregnancy (see also Abalovich et al. [54]) (Fig. 5.3).

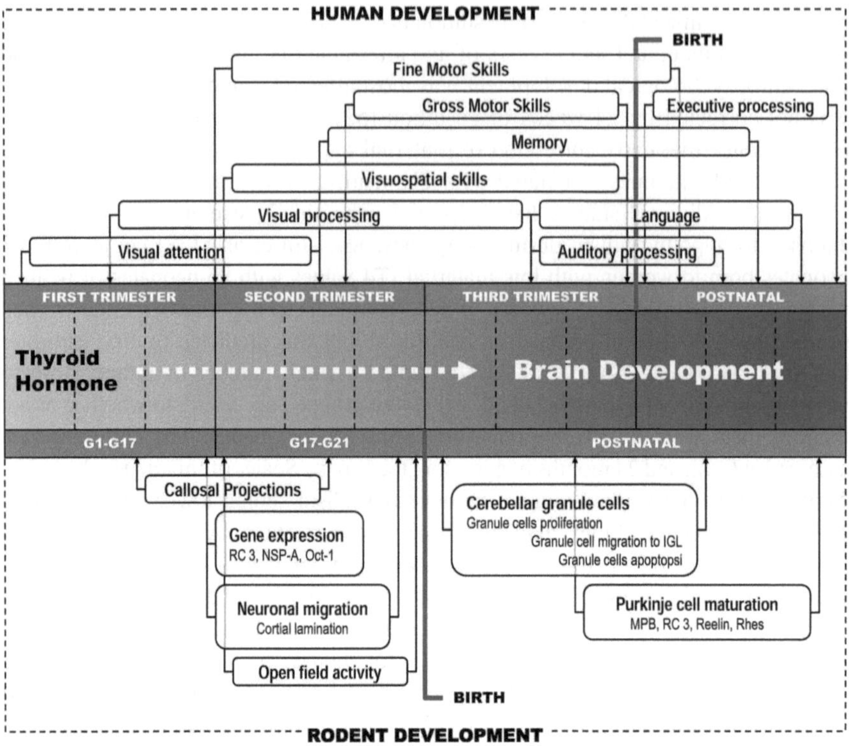

Fig. 5.3 Differential timing of effects of TH insufficiency in humans (*upper panel*) and rodents (*lower panel*). Based on studies of TH deficiency in humans limited to early prenatal (maternal hypothyroidism), late prenatal (premature birth), and early postnatal (congenital hypothyroidism) periods. TH insufficiency during fetal development exerts greater effects on cortical development, whereas postnatal hypothyroidism exerts greater effects on cerebellar development. Reprinted from Zoeller and Rovet [80], with permission from Wiley–Blackwell Publishing

Overall, the studies of Haddow et al. [4] and Pop et al. [5] have made significant contributions to our current understanding of the long-term effects of maternal hypothyroidism on child development. Because of their findings, significant debate has arisen in both research and medical communities regarding the implementation of universal TH screening programs to identify women with either clinical or subclinical forms of hypothyroidism and whether routine thyroxine-replacement treatment should be prescribed for women diagnosed with subclinical hypothyroidism [47, 54, 144]. This debate reflects a lack of consensus on when best to conduct such screening (i.e., before or during pregnancy), which hormone to screen for (i.e., total T4, free T4, TSH), how to screen (e.g., filter paper blood samples, venous serum sampling), and whether there is a benefit of treatment for the offspring of women with subclinical hypothyroidism [145].

At present, the long-term results from the one major randomized control study from Wales, known as the Controlled Antenatal Thyroid Screening study (CATS) trial, are incomplete [146]. Despite the research by Morreale de Escobar et al. [20]

showing that low maternal T4 levels during pregnancy pose a significant risk for child neuropsychological development, current obstetric practices do not involve widespread screening for thyroid disease unless the patient has risk factors (e.g., having thyroid disease or a family history of thyroid disorders) or presents with symptoms of hypothyroidism [54, 62, 147]. Although current guidelines recommend careful case-finding, this technique is far from perfect and may miss or improperly classify many cases [145].

Does Maternal Hypothyroidism Contribute to Childhood Developmental Disorders?

At present, very little is known about the contribution of maternal hypothyroidism during pregnancy to subsequent developmental disorders of childhood. Nevertheless, several studies report data suggesting an increased incidence of attention difficulties [4, 148], and a recent paper also postulates that maternal hypothyroidism during pregnancy, particularly low T3 levels during early periods of cell migration in the fetal brain, may be associated with an increased prevalence of autism in the general population [149].

In addition to detailed analyses on IQ in the study by Haddow et al. [4], interesting data were reported on two aspects of attention. Children of the entire group of women with elevated TSH levels had significantly lower scores than controls on a computerized test of visual attention known as the Continuous Performance Test, and they had lower scores (at a trend level) on an index of auditory attention. Moreover, when the children of hypothyroid women were stratified according to whether or not mothers received treatment, the offspring of untreated mothers scored lower than controls on both aspects of attention [4]. Thus, these results suggest a possible association between maternal hypothyroidism and attention problems in the progeny. Indeed, Vermiglio et al. [148] observed that children born in a moderately iodine-deficient area of Sicily had a substantially increased risk of attention deficit hyperactivity disorder (ADHD) compared with children born to women from an area of iodine sufficiency and that ADHD was strongly associated with maternal hypothyroxinemia in early gestation [148].

Recent work by Román [149] has examined whether a possible relationship could exist between autism and maternal hypothyroidism due to iodine deficiency, as well as naturally occurring environmental goitrogens (kale, sweet potatoes, cassava) and environmental thyroid disruptors, such as percholorates, polycholorinated biphenyls, phthalates, common herbicides (such as acetochlor), and thyiocyanate in tobacco smoke. Román [149] outlines similarities in the neuropathology of autism and animal models of congenital hypothyroidism, particularly in disrupted reelin and Dab1 gene expression, resulting in abnormal neuronal migration. In addition, Román [149] also reports that the risk of autism is doubled in individuals with a family history of autoimmune thyroiditis. However, no studies to date have directly investigated a link between autism and maternal thyroid function during the first or second trimesters of pregnancy, thus further research is required. Of interest in our own clinical experience, autism has been seen in several children born to women whose hyperthyroidism was overtreated (thereby shutting down both maternal and fetal systems) in different trimesters of pregnancy, as well as in children with delayed treatment of congenital hypothyroidism.

Conclusions

This review has provided convincing evidence showing that thyroid hormones of maternal origin are needed throughout gestation for proper brain development and if insufficient, long-lasting effects are observed in the progeny. Moreover, the specific effects will differ depending on when the maternal TH insufficiency occurs during pregnancy. Given the results of these studies, it is recommended that all pregnant women with overt hypothyroidism should be continuously monitored and adequately treated. In addition, there is a need to correct thyroid function in women with other gestational TH insufficiencies through proper screening and treatment. However, the debate on whether population screening for maternal TH insufficiency is necessary has not been resolved and awaits the results of current clinical trials in progress.

Clearly, further research is needed to determine exactly how maternal thyroid insufficiency affects childhood outcomes, particularly at older ages when results are permanent, and with technologies such as magnetic resonance imaging (MRI) and functional MRI, which will permit direct assessment of the impact of early TH loss on the developing human brain. Finally, there is a need for further information on whether and how maternal TH insufficiency during pregnancy may contribute to various neurodevelopmental disorders of childhood, especially those in which the incidence is steadily increasing.

References

1. Bernal J, Nunez J (1995) Thyroid hormones and brain development. Eur J Endocrinol 133:390–398.
2. Koibuchi N, Chin WW (2000) Thyroid hormone action and brain development. Trends Endocrinol Metab 11:123–128.
3. Anderson GW (2001) Thyroid hormones and the brain. Front Neuroendocrinol 22:1–17.
4. Haddow JE, Palomaki GE, Allan WC, Williams JR, Knight GJ, Gagnon J, O'Heir CE, Mitchell ML, Hermos RJ, Waisbren SE, Faix JD, Klein RZ (1999) Maternal thyroid deficiency during pregnancy and subsequent neuropsychological development of the child. N Engl J Med 341:549–555.
5. Pop VJ, Kuijpens JL, van Baar AL, Verkerk G, van Son MM, de Vijlder JJ, Vulsma T, Wiersinga WM, Drexhage HA, Vader HL (1999) Low maternal free thyroxine concentrations during early pregnancy are associated with impaired psychomotor development in infancy. Clin Endocrinol 50:149–155.

6. Bettendorf M (2002) Thyroid disorders in children from birth to adolescence. Eur J Nucl Med Mol Imaging 29:S429–S446.
7. de Morreale Escobar G, Obregón MJ, del Escobar Rey F (2004) Maternal thyroid hormones early in pregnancy and fetal brain development. Best Pract Res Clin Endocrinol Metab 18:225–248.
8. Calvo RM, Jauniaux E, Gulbis B, Asuncion M, Gervy C, Contempré B, de Morreale Escobar G (2002) Fetal tissues are exposed to biologically relevant free thyroxine concentrations during early phases of development. J Clin Endocrinol Metab 87:1768–1777.
9. de Morreale Escobar G (2001) The role of thyroid hormone in fetal neurodevelopment. J Pediatr Endocrinol Metab 14:1453–1462.
10. Anderson GW, Schoonover CM, Jones SA (2003) Control of thyroid hormone action in the developing rat brain. Thyroid 13:1039–1056.
11. Burrow GN, Fisher DA, Larsen PR (1994) Maternal and fetal thyroid function. N Engl J Med 331:1072–1078.
12. Vulsma T, Gons MH, de Vijlder JJ (1989) Maternal-fetal transfer of thyroxine in congenital hypothyroidism due to a total organification defect or thyroid agenesis. N Engl J Med 321:13–16.
13. Oppenheimer JH, Schwartz HL (1997) Molecular basis of thyroid hormone-dependent brain development. Endocr Rev 18:462–475.
14. Costa A, Arisio R, Benedetto C, Bertino E, Fabris C, Giraudi G, Marozio L, Maulà V, Pagliano M, Testori O, Zoppetti G (1991) Thyroid hormones in tissues from human embryos and fetuses. J Endocrinol Invest 14:559–568.
15. Contempré B, Jauniaux E, Calvo R, Jurkovic D, Campbell S, de Morreale Escobar G (1993) Detection of thyroid hormones in human embryonic cavities during the first trimester of pregnancy. J Clin Endocrinol Metab 77:1719–1722.
16. Kilby MD, Verhaeg J, Gittoes N, Somerset DA, Clark PMS, Franklyn JA (1998) Circulating thyroid hormone concentrations and placental thyroid hormone receptor expression in normal human pregnancy and pregnancy complicated by intrauterine growth restriction (IUGR). J Clin Endocrinol Metab 83:2964–2971.
17. Thorpe-Beeston JG, Nicolaides KH, Felton CV, Butler J, McGregor AM (1991) Maturation of the secretion of thyroid hormones and thyroid-stimulating hormone in the fetus. N Engl J Med 324:532–536.
18. Blazer S, Moreh-Waterman M, Miller-Lotan R, Tamir A, Hochberg Z (2003) Maternal hypothyroidism may affect fetal growth and neonatal thyroid function. Obstet Gynecol 102:232–241.
19. Klein RZ, Mitchell ML (2002) Maternal hypothyroidism and cognitive development of the offspring. Curr Opin Pediatr 14:443–446.
20. de Morreale Escobar G, Obregón MJ, del Escobar Rey F (2000) Is neurodevelopment related to maternal hypothyroidism or to maternal hypothyroxinemia? J Clin Endocrinol Metab 85:3975–3987.
21. Casey BM, Dashe JS, Wells CE, McIntire DD, Byrd W, Leveno KJ, Cunningham FG (2005) Subclinical hypothyroidism and pregnancy outcomes. Obstet Gynecol 105:239–245.
22. Leung AS, Millar LK, Koonings PP, Montoro M, Mestman JH (1993) Perinatal outcome in hypothyroid pregnancies. Obstet Gynecol 81:349–353.
23. Stagnaro-Green A, Chen X, Bogden JD, Davies TF, Scholl TO (2005) The thyroid and pregnancy: a novel risk factor for very preterm delivery. Thyroid 15:351–357.
24. Howdeshell KL (2002) A model of the development of the brain as a construct of the thyroid system. Environ Health Perspect 110:337–348.
25. Williams GR (2008) Neurodevelopmental and neurophysiological actions of thyroid hormone. J Neuroendocrinol 20:784–794.
26. Zimmermann MB (2009) Iodine deficiency. Endocr Rev 30:376–408.
27. Surks MI, Schadlow AR, Stock JM, Oppenheimer JH (1973) Determination of iodothyronine absorption and conversion of L-thyroxine (T4) to L-triiodothyronine (T3) using turnover rate techniques. J Clin Invest 52:805–811.

28. Flamant F, Samarut J (2003) Thyroid hormone receptors: lessons from knockout and knock-in mutant mice. Trends Endocrinol Metab 14:85–90.
29. Larson PR, Silva JE, Kaplan MM (1981) Relationships between circulating and intracellular thyroid hormones: physiological and clinical implications. Endocr Rev 2:87–102.
30. Kester MH, de Martinez Mena R, Obregon MJ, Marinkovic D, Howatson A, Visser TJ, Hume R, de Morreale Escobar G (2004) Iodothyronine levels in the human developing brain: major regulatory roles of iodothyronine deiodinases in different areas. J Clin Endocrinol Metab 89:3117–3128.
31. Bianco AC, Salvatore D, Gereben B, Berry MJ, Larsen PR (2002) Biochemistry, cellular and molecular biology, and physiological roles of the iodothyronine selenodeiodinases. Endocr Rev 23:38–89.
32. Bernal J (2007) Thyroid hormone receptors in brain development and function. Nat Clin Pract Endocrinol Metab 3:249–259.
33. Daras VM, Hume R, Visser TJ (1999) Regulation of thyroid hormone metabolism during fetal development. Mol Cell Endocrinol 151:37–47.
34. Leonard JL, Köhrle J (1996) Intracellular pathways of iodothyronine metabolism. In: Braverman LE, Utiger RD (eds) The thyroid. Lippincott-Raven, Philadelphia, pp 125–161.
35. Guadaño-Ferraz A, Escámez MJ, Rausell E, Bernal J (1999) Expression of type 2 iodothyronine deiodinase in hypothyroid rat brain indicates an important role of thyroid hormone in the development of specific primary sensory systems. J Neurosci 19:3430–3439.
36. Bernal J (2005) Thyroid hormones and brain development. Vitam Horm 71:95–122.
37. Morte B, Manzano J, Scanlan T, Vennström B, Bernal J (2002) Deletion of the thyroid hormone receptor alpha 1 prevents the structural alterations of the cerebellum induced by hypothyroidism. Proc Natl Acad Sci U S A 99:3985–3989.
38. Chan SY, Vasilopoulou E, Kilby MD (2009) The role of the placenta in thyroid hormone delivery to the fetus. Nat Clin Pract Endocrinol Metab 5:45–54.
39. Glinoer D (1997) The regulation of thyroid function in pregnancy: pathways of endocrine adaptation from physiology to pathology. Endocr Rev 18:404–433.
40. Shan ZY, Chen YY, Teng WP, Yu XH, Li CY, Zhou WW, Gao B, Zhou JR, Ding B, Ma Y, Wu Y, Liu Q, Xu H, Liu W, Li J, Wang WW, Li YB, Fan CL, Wang H, Guo R, Zhang HM (2009) A study for maternal thyroid hormone deficiency during the first half of pregnancy in China. Eur J Clin Invest 39:37–42.
41. Glinoer D (2004) The regulation of thyroid function during normal pregnancy: importance of the iodine nutrition status. Best Pract Res Clin Endocrinol Metab 18:133–152.
42. Ausó E, Lavado-Autric R, Cuevas E, del Escobar Rey F, de Morreale Escobar G, Berbel P (2004) A moderate and transient deficiency of maternal thyroid function at the beginning of fetal neocorticogenesis alters neuronal migration. Endocrinology 145:4037–4047.
43. Santini F, Chiovato L, Ghirri P, Lapi P, Mammoli C, Montanelli L, Scartabelli G, Ceccarini G, Coccoli L, Chopra IJ, Boldrini A, Pinchera A (1999) Serum iodothyronines in the human fetus and the newborn: evidence for an important role of placenta in fetal thyroid hormone homeostasis. J Clin Endocrinol Metab 84:493–498.
44. Bernal J, Pekonen F (1984) Ontogenesis of the nuclear 3, 5, 3′-triiodothyronine receptor in the human fetal brain. Endocrinology 114:677–679.
45. Ferreiro B, Bernal J, Goodyer CG, Branchard CL (1988) Estimation of nuclear thyroid hormone receptor saturation in human fetal brain and lung during early gestation. J Clin Endocrinol Metab 67:853–856.
46. Dowling AL, Martz GU, Leonard JL, Zoeller RT (2000) Acute changes in maternal thyroid hormone induce rapid and transient changes in gene expression in fetal rat brain. J Neurosci 20:2255–2265.
47. Chan S, Rovet J (2003) Thyroid hormones in fetal central nervous system development. Fetal Matern Med Rev 13:177–208.
48. Alexander EK, Marqusee E, Lawrence J, Jarolim P, Fischer GA, Larsen PR (2004) Timing and magnitude of increases in levothyroxine requirements during pregnancy in women with hypothyroidism. N Engl J Med 351:241–249.

49. Glinoer D, Delange F (2000) The potential repercussions of maternal, fetal, and neonatal hypothyroxinemia on the progeny. Thyroid 10:871–887.
50. Casey BM (2006) Subclinical hypothyroidism and pregnancy. Obstet Gynecol Surv 61:415–420.
51. Allan WC, Haddow JE, Palomaki GE, Williams JR, Mitchell ML, Hermos RJ, Faix JD, Klein RZ (2000) Maternal thyroid deficiency and pregnancy complications: implications for population screening. J Med Screen 7:127–130.
52. Pop VJ, Brouwers EP, Vader HL, Vulsma T, van Baar AL, de Vijlder JJ (2003) Maternal hypothyroxinaemia during early pregnancy and subsequent child development: a 3-year follow-up study. Clin Endocrinol 59:282–288.
53. Glinoer D (1998) The systematic screening and management of hypothyroidism and hyperthyroidism during pregnancy. Trends Endocrinol Metab 9:403–411.
54. Abalovich M, Amino N, Barbour LA, Cobin RH, de Groot LJ, Glinoer D, Mandel SJ, Stagnaro-Green A (2007) Management of thyroid dysfunction during pregnancy and postpartum: an endocrine society clinical practice guideline. J Clin Endocrinol Metab 92:S1–S47.
55. Kamijo K, Saito T, Sato M, Yachi A, Mukai A, Fukusi M et al (1990) Transient subclincial hypothyroidism in early pregnancy. Endocrinol Jpn 37:397–403.
56. Utiger RD (1999) Maternal hypothyroidism and fetal development. N Engl J Med 341:601–602.
57. Klein RZ, Sargent JD, Larsen PR, Waisbren SE, Haddow JE, Mitchell ML (2001) Relation of severity of maternal hypothyroidism to cognitive development of offspring. J Med Screen 8:18–20.
58. Panesar NS, Li CY, Rogers MS (2001) Reference intervals for thyroid hormones in pregnant Chinese women. Ann Clin Biochem 38:329–332.
59. Mandel SJ (2004) Hypothyroidism and chronic autoimmune thyroiditis in the pregnant state: maternal aspects. Best Pract Res Clin Endocrinol Metab 18:213–224.
60. Soldin OP, Tractenberg RE, Hollowell JG, Jonklaas J, Janicic N, Soldin SJ (2004) Trimester-specific changes in maternal thyroid hormone, thyrotropin, and thyroglobulin concentrations during gestation: trends and associations across trimesters in iodine sufficiency. Thyroid 14:1084–1090.
61. Becker DV, Bigos ST, Gaitan E, Morris JC, Rallison ML, Spencer CA, Sugawara M, Van Middlesworth L, Wartofsky L (1993) Optimal use of blood tests for assessment of thyroid function. JAMA 269:2736–2737.
62. Ladenson PW, Singer PA, Ain KB, Bagchi N, Bigos ST, Levy EG, Smith SA, Daniels GH, Cohen HD (2000) American thyroid association guidelines for detection of thyroid dysfunction. Arch Intern Med 160:1573–1575.
63. Davis LE, Leveno KJ, Cunningham FG (1988) Hypothyroidism complicating pregnancy. Obstet Gynecol 72:108–112.
64. Idris I, Srinivasan R, Simm A, Page RC (2005) Maternal hypothyroidism in early and late gestation: effects on neonatal and obstetric outcome. Clin Endocrinol 63:560–565.
65. Klein RZ, Haddow JE, Faix JD, Brown RS, Hermos RJ, Pulkkinen A, Mitchell ML (1991) Prevalence of thyroid deficiency in pregnant women. Clin Endocrinol 35:41–46.
66. Lazarus JH (2002) Epidemiology and prevention of thyroid disease in pregnancy. Thyroid 12:861–865.
67. Woeber KA (1997) Subclinical thyroid dysfunction. Arch Intern Med 157:1065–1068.
68. Oken E, Braverman LE, Platek D, Mitchell ML, Lee SL, Pearce EN (2009) Neonatal thyroxine, maternal thyroid function, and child cognition. J Clin Endocrinol Metal 94:497–503.
69. Hollowell JG, Staehling NW, Flanders WD, Hannon WH, Gunter EW, Spencer CA, Braverman LE, Serum TSH (2002) T(4), and thyroid antibodies in the United States population (1988 to 1994): National Health and Nutrition Examination Survey (NHANES III). J Clin Endocrinol Metab 87:489–499.
70. Leung AM, Pearce EN, Braverman LE (2009) Iodine content of prenatal multivitamins in the United States. N Engl J Med 360:939–940.

71. Becker DV, Braverman LE, Delange F, Franklyn JA, Hollowell JG, Lamm SH, Mitchell ML, Pearce E, Robbins J, Rovet J (2006) Iodine supplementation for pregnancy and lactation – United States and Canada: recommendations of the American Thyroid Association. Thyroid 16:949–951.
72. Gillam MP, Kopp P (2001) Genetic regulation of thyroid development. Curr Opin Pediatr 13:358–363.
73. Thorpe-Beeston JG, Nicolaides KH, McGregor AM (1992) Fetal thyroid function. Thyroid 2:207–217.
74. van Wassenaer AG, Kok JH (2004) Hypothyroxinaemia and thyroid function after preterm birth. Semin Neonatol 9:3–11.
75. Ballabio M, Nicolini U, Jowett T, Ruiz de Elvira MC, Ekins RP, Rodeck CH (1989) Maturation of thyroid function in normal human foetuses. Clin Endocrinol 31:565–571.
76. Kilby MD, Gittoes N, McCabe C, Verhaeg J, Franklyn JA (2000) Expression of thyroid receptor isoforms in the human fetal central nervous system and the effects of intrauterine growth restriction. Clin Endocrinol 53:469–477.
77. Chan S, Kachilele S, McCabe CJ, Tannahill LA, Boelaert K, Gittoes NJ, Visser TJ, Franklyn JA, Kilby MD (2002) Early expression of thyroid hormone deoidinases and receptors in human fetal cerebral cortex. Brain Res Dev Brain Res 138:109–116.
78. Hosoda R, Nakayama K, Kato-Negishi M, Kawahara M, Muramoto K, Kuroda Y (2003) Thyroid hormone enhances the formation of synapses between cultured neurons of rat cerebral cortex. Cell Mol Neurobiol 23:895–906.
79. Obregón MJ, Calvo RM, Del Rey FE, de Escobar GM (2007) Ontogenesis of thyroid function and interactions with maternal function. Endocr Dev 10:86–89.
80. Zoeller RT, Rovet J (2004) Timing of thyroid hormone action in the developing brain: clinical observations and experimental findings. J Neuroendocrinol 16:809–818.
81. Lavado-Autric R, Ausó E, García-Velasco JV, Arufe Mdel C, del Escobar Rey F, Berbel P, de Morreale Escobar G (2003) Early maternal hypothyroxinemia alters histogenesis and cerebral cortex cytoarchitecture of the progeny. J Clin Invest 111:954–957.
82. Cuevas E, Ausó E, Telefont M, de Morreale Escobar G, Sotelo C, Berbel P (2005) Transient maternal hypothyroxinemia at onset of corticogenesis alters tangential migration of medial ganglionic eminence-derived neurons. Eur J Neurosci 22:541–551.
83. Goodman JH, Gilbert ME (2007) Modest thyroid hormone insufficiency during development induces a cellular malformation in the corpus callosum: a model of cortical dysplasia. Endocrinology 148:2593–2597.
84. Porterfield SP, Hendrich CE (1993) The role of thyroid hormones in prenatal and neonatal neurological development – current perspectives. Endocr Rev 14:94–106.
85. Chan S, Kilby MD (2000) Thyroid hormone and central nervous system development. J Endocrinol 165:1–8.
86. Sui L, Gilbert ME (2003) Pre- and postnatal propylthiouracil-induced hypothyroidism impairs synaptic transmission and plasticity in area CA1 of the neonatal rat hippocampus. Endocrinology 144:4195–4203.
87. Evans IM, Sinha AK, Pickard MR, Edwards PR, Leonard AJ, Ekins RP (1999) Maternal hypothyroxinemia disrupts neurotransmitter metabolic enzymes in developing brain. J Endocrinol 161:273–279.
88. Vara H, Martínez B, Santos A, Colino A (2002) Thyroid hormone regulates neurotransmitter release in neonatal rat hippocampus. Neuroscience 110:19–28.
89. Fäldt R, Passant U, Nilsson K, Wattmo C, Gustafson L (1996) Prevalence of thyroid hormone abnormalities in elderly patients with symptoms of organic brain disease. Aging 8:347–353.
90. Mafrica F, Fodale V (2008) Thyroid function, Alzheimer's disease and postoperative cognitive dysfunction: a tale of dangerous liaisons? J Alzheimers Dis 14:95–105.
91. Bunevicius R (2009) Thyroid disorders in mental patients. Curr Opin Psychiatry 22:391–395.
92. Forrest D, Reh TA, Rüsch A (2002) Neurodevelopmental control by thyroid hormone receptors. Curr Opin Neurobiol 12:49–56.

93. Thompson CC, Potter GB (2000) Thyroid hormone action in neural development. Cereb Cortex 10:939–945.
94. Dugbartey AT (1998) Neurocognitive aspects of hypothyroidism. Arch Intern Med 158:1413–1418.
95. Weller A, Rozin A, Rigler O, Sack J (1996) Neurobehavioral development of neonatal rats after in-utero hypothyroxinemia: efficacy of prenatal thyroxine treatment. Early Hum Dev 46:63–76.
96. Dowling AL, Iannacone EA, Zoeller RT (2001) Maternal hypothyroidism selectively affects the expression of neuroendocrine-specific protein A messenger ribonucleic acid in the proliferative zone of the fetal rat brain cortex. Endocrinology 142:390–399.
97. Schwartz HL (1983) Effect of thyroid hormone on growth and development. In: Oppenheimer JH, Samuels HH (eds) Molecular basis of thyroid hormone action. Academic Press, New York, NY, pp 413–444.
98. Chan SY, Franklyn JA, Kilby MD (2005) Maternal thyroid hormones and fetal brain development. Curr Opin Endocrinol Diabetes 12:23–30.
99. Sidman RL, Rakic P (1973) Neuronal migration, with special reference to developing human brain: a review. Brain Res 101:465–470.
100. Lucio RA, García JV, Ramón Cerezo J, Pacheco P, Innocenti GM, Berbel P (1997) The development of auditory callosal connections in normal and hypothyroid rats. Cereb Cortex 7:303–316.
101. Naranyanan CH, Narayanan Y (1985) Cell formation in the motor nucleus and mesencephalic nucleus of the trigeminal nerve of rats made hypothyroid by propylthiouracil. Exp Brain Res 59:257–266.
102. Berbel P, Guadaño-Ferraz A, Angulo A, Ramón Cerezo J (1994) Role of thyroid hormones in the maturation of interhemispheric connections in rats. Behav Brain Res 64:9–14.
103. Gravel C, Sasseville R, Hawkes R (1990) Maturation of the corpus callosum of the rat: II. Influence of thyroid hormones on the number and maturation of axons. J Comp Neurol 291:147–161.
104. Guadaño Ferraz A, del Escobar Rey F, de Morreale Escobar G, Innocenti GM, Berbel P (1994) The development of the anterior commissure in normal and hypothyroid rats. Brain Res Dev Brain Res 81:293–308.
105. Sharlin DS, Tighe D, Gilbert ME, Zoeller RT (2008) The balance between oligodendrocyte and astrocyte production in major white matter tracts is linearly related to serum total thyroxine. Endocrinology 149:2527–2536.
106. Square LR (1992) Memory and the hippocampus: a synthesis from findings with rats, monkeys, and humans. Psychol Rev 99:195–231.
107. Jones-Gotman M, Zatorre RJ, Olivier A, Andermann F, Cendes F, Staunton H, McMackin D, Siegel AM, Wieser HG (1997) Learning and retention of words and designs following excision from medial or lateral temporal-lobe structures. Neuropsychologia 35:963–973.
108. Madeira MD, Sousa N, Lima-Andrade MT, Calheiros F, Cadete-Leite A, Paula-Barbosa MM (1992) Selective vulnerability of the hippocampal pyramidal neurons to hypothyroidism in male and female rats. J Comp Neurol 322:501–518.
109. Taylor MA, Swant J, Wagner JJ, Fisher JW, Ferguson DC (2008) Lower thyroid compensatory reserve of rat pups after maternal hypothyroidism: correlation of thyroid, hepatic, and cerebrocortical biomarkers with hippocampal neurophysiology. Endocrinology 149:3521–3530.
110. Rami A, Patel AJ, Rabié A (1986) Thyroid hormone and development of the rat hippocampus: Morphological alterations in granule and pyramidal cells. Neuroscience 19:1217–1226.
111. Gilbert ME, Paczowski C (2003) Propylthiouracil (PTU)-induced hypothyroidism in the developing rat impairs synaptic transmission and plasticity in the dentate gyrus of the adult hippocampus. Brain Res Dev Brain Res 145:19–29.
112. Opazo MC, Gianini A, Pancetti F, Azkcona G, Alarcón L, Lizana R, Noches V, Gonzalez PA, Marassi MP, Mora S, Rosenthal D, Eugenin E, Naranjo D, Bueno SM, Kalergis AM, Riedel CA (2008) Maternal hypothyroxinemia impairs spatial learning and synaptic nature and function in the offspring. Endocrinology 149:5097–5106.

113. Gilbert ME, Sui L (2006) Dose-dependent reductions in spatial learning and synaptic function in the dentate gyrus of adult rats following developmental thyroid hormone insufficiency. Brain Res 1069:10–22.
114. Heuer H, Mason CA (2003) Thyroid hormone induces cerebellar Purkinje cell dendritic development via the thyroid hormone receptor alpha1. J Neurosci 23:10604–10612.
115. Nicolson JL, Altman J (1972) The effects of early hypo- and hyperthyroidism on the development of rat cerebellar cortex. I. Cell proliferation and differentiation. Brain Res 44:13–23.
116. Nicolson JL, Altman J (1972) The effects of early hypo- and hyperthyroidism on the development of rat cerebellar cortex. II. Synaptogenesis in the molecular layer. Brain Res 44:25–36.
117. Gravel C, Hawkes R (1990) Maturation of the corpus callosum of the rat: I. Influence of thyroid hormones on the topography of callosal projections. J Comp Neurol 291:128–146.
118. Li CP, Olavarria JF, Greger BE (1995) Occipital cortico-pyramidal projection in hypothyroid rats. Brain Res Dev Brain Res 89:227–234.
119. Ruiz-Marcos A, Sanchez-Toscano F, Obregon MJ, Escobar del Rey F, Morreale de Escobar G (1982) Thyroxine treatment and recovery of hypothyroidism-induced pyramidal cell damage. Brain Res 239:559–574.
120. Fagge CH (1871) On sporadic cretinism occurring in England. Med Chir Trans 54:155.
121. Merke H (1984) History and iconography of endemic goiter and cretinism. Hans Huber, Bern.
122. Flint A (1873) The physiology of Man. D. Appleton and Company, New York.
123. McCarrison R (1917) The thyroid gland in health and disease. Balliére, Tindall & Cox, London.
124. Hunziker-Shild H (1915) Der Kropf, eine Anpassung an Jodarme Nahrung. Verlage A. Franke, Berne.
125. Fisher DA, Hobel CJ, Garza R, Pierce CA (1970) Thyroid function in the preterm fetus. Pediatrics 46:208–216.
126. Greenberg AH, Czernichow P, Reba RC, Tyson J, Blizzard RM (1970) Observations on the maturation of thyroid function in early fetal life. J Clin Invest 49:1790–1803.
127. Harington CR, Barger G (1927) Chemistry of thyroxine: constitution and synthesis of thyroxine. Biochem J 21:169–183.
128. Gross J, Pitt-Rivers R (1952) The identification of 3:5:3′-L-triiodothyronine in human plasma. Lancet 1:439–441.
129. Choufoer JC, Vanrhihn M, Querido A (1965) Endemic goiter in western New Guinea. II. Clinical picture, incidence and pathogenesis of endemic cretinism. J Clin Endocrinol Metab 25:385–402.
130. Pharoah PO, Connolly KJ, Ekins RP, Harding AG (1984) Maternal thyroid hormone levels in pregnancy and the subsequent cognitive and motor performance of the children. Clin Endocrinol (Oxf) 21:265–270.
131. Roti E, Gnudi A, Braverman LE (1983) The placental transport, synthesis and metabolism of hormones and drugs which affect thyroid function. Endocr Rev 4:131–149.
132. Man EB, Brown JF, Serunian SA (1991) Maternal hypothyroxinemia: psychoneurological deficits of progeny. Ann Clin Lab Sci 21:227–239.
133. Smallridge RC, Ladenson PW (2001) Hypothyroidism in pregnancy: consequences to neonatal health. J Clin Endocrinol Metab 86:2349–2353.
134. Wasserstrum N, Anania CA (1995) Perinatal consequences of maternal hypothyroidism in early pregnancy and inadequate replacement. Clin Endocrinol (Oxf) 42:353–358.
135. Casey BM, Dashe JS, Spong CY, McIntire DD, Leveno KJ, Cunningham GF (2007) Perinatal significance of isolated maternal hypothyroxinemia identified in the first half of pregnancy. Obstet Gynecol 109:1129–1135.
136. Negro R, Mangieri T, Coppola L, Presicce G, Casavola EC, Gismondi R, Locorotondo G, Caroli P, Pezzarossa A, Dazzi D, Hassan H (2005) Levothyroxine treatment in thyroid peroxidase antibody-positive women undergoing assisted reproduction technologies: a prospective study. Hum Reprod 20:1529–1533.

137. Dussault JH, Fisher DA (1999) Thyroid function in mothers of hypothyroid newborns. Obstet Gynecol 93:15–20.
138. Kooistra L, Crawford S, van Baar AL, Brouwers EP, Pop VJ (2006) Neonatal effects of maternal hypothyroxinemia during early pregnancy. Pediatrics 117:161–167.
139. Smit BJ, Kok JH, Vulsma T, Briët JM, Boer K, Wiersinga WM (2000) Neurologic development of the newborn and young child in relation to maternal thyroid function. Acta Paediatr 89:291–295.
140. Mirabella G, Feig D, Astzalos E, Perlman K, Rovet JF (2000) The effect of abnormal intrauterine thyroid hormone economies on infant cognitive abilities. J Pediatr Endocrinol Metab 13:191–194.
141. Mirabella G, Westall CA, Asztalos E, Perlman K, Koren G, Rovet J (2005) Development of contrast sensitivity in infants with prenatal and neonatal thyroid hormone insufficiencies. Pediatr Res 57:902–907.
142. Matsuura N, Konishi J (1990) Transient hypothyroidism in infants born to mothers with chronic thyroiditis – a nationwide study of twenty-three cases. The Transient Hypothyroidism Study Group. Endocrinol Jpn 37:369–379.
143. Liu H, Momotani N, Noh JY, Ishikawa N, Takebe K, Ito K (1994) Maternal hypothyroidism during early pregnancy and intellectual development of the progeny. Arch Intern Med 154:785–787.
144. Gharib H, Cobin R, Dickey R (1999) Subclinical hypothyroidism during pregnancy: position statement from the American Association of Clinical Endocrinologists. Endocr Pract 5:367–368.
145. Vaidya B, Anthony S, Bilous M, Shields B, Drury J, Hutchison S, Bilous R (2007) Detection of thyroid dysfunction in early pregnancy: universal screening or targeted high-risk case finding? J Clin Endocrinol Met 92:203–207.
146. Lazarus JH, Premawardhana LD (2005) Screening for thyroid disease in pregnancy. J Clin Pathol 58:449–452.
147. Gyamfi C, Wapner RJ, D'Alton ME (2009) Thyroid dysfunction in pregnancy: the basic science and clinical evidence surrounding the controversy in management. Obstet Gynecol 113:702–707.
148. Vermiglio F, Lo Presti VP, Moleti M, Sidoti M, Tortorell G, Scaffidi G, Castagna MG, Mattina F, Violi MA, Crisa A, Artemisia A, Trimarchi F (2004) Attention deficit and hyperactivity disorders in the offspring of mothers exposed to mild-moderate iodine deficiency: a possible novel iodine deficiency disorder in developed countries. J Clin Endocrinol Met 89:6054–6060.
149. Román GC (2007) Autism: transient in utero hypothyroxinemia related to maternal flavonoid ingestion during pregnancy and to other environmental antithyroid agents. J Neurol Sci 262:15–26.
150. de Morreale Escobar GM, Obregón MJ, del Rey FE (2007) Iodine deficiency and brain development in the first half of pregnancy. Public Health Nutr 10:1554–1570.

Chapter 6
Obstetric Factors Related to Perinatal Brain Injury

Christopher S. Ennen and Ernest M. Graham

Keywords Hypoxic ischemic encephalopathy • Metabolic acidosis • Cardiotocography • Sentinel event • Periventricular leukomalacia

Introduction

Labor and delivery are the culminating events of the maternal–fetal relationship. Although the outcome for mother and child is favorable in the majority of cases, events may occur during parturition that affect the future neurological status of the fetus. In this chapter, we will review the definitions, incidence, and pathophysiology of neonatal encephalopathy and cerebral palsy, examine the impact that labor and delivery have on the fetus, review the methods used to evaluate for fetal compromise during labor and neonatal markers that predict brain injury, and examine the potential interventions to reduce the risk of perinatal brain injury.

Definitions and Epidemiology

Historically, the process of birth has been deemed the likely cause of most postpartum neurological injuries in infants [1]. "Birth asphyxia" has been a commonly used term, often without a clear definition. As recently as the 1970s, it was assumed that intrapartum events causing hypoxic-ischemic injury were responsible for half of perinatal morbidity and mortality [2]. Early studies varied in their definition of cases, making comparisons difficult.

E.M. Graham (✉)
Maternal-Fetal Medicine Division, Johns Hopkins Hospital, Phipps 228, 600 N. Wolfe St., Baltimore, MD, 21287, USA
e-mail: egrahaa@jhmi.edu

A.W. Zimmerman and S.L. Connors (eds.), *Maternal Influences on Fetal Neurodevelopment: Clinical and Research Aspects*,
DOI 10.1007/978-1-60327-921-5_6, © Springer Science+Business Media, LLC 2010

In 2003, the American College of Obstetricians and Gynecologists (ACOG), in conjunction with the American Academy of Pediatrics (AAP), published "Neonatal Encephalopathy and Cerebral Palsy: Defining the Pathogenesis and Pathophysiology" [3]. They defined neonatal encephalopathy as a clinical condition including a "combination of abnormal consciousness, tone and reflexes, feeding, respiration, or seizures and can result from myriad conditions." Cerebral palsy is defined as a nonprogressive "chronic disability of central nervous system origin characterized by aberrant control of movement and posture." They also state that any perinatal brain injury sufficient to cause cerebral palsy must "progress through neonatal encephalopathy."

In the 1980s and 1990s, many studies demonstrated the rate of intrapartum injury to be much lower than previously thought [4]. A recent systematic review of the role of intrapartum hypoxia-ischemia as a cause of neonatal encephalopathy concluded that 3.7 per 1,000 term neonates are born with evidence of intrapartum hypoxia (based on an umbilical artery pH of <7.0) with 23.1% of these infants having neurological morbidity or mortality [5]. The incidence of neonatal hypoxic-ischemic encephalopathy was 2.5 per 1,000 live births and 14.5% of cases of cerebral palsy were associated with intrapartum hypoxia.

The ACOG task force authoring the previously discussed monograph concluded that current scientific evidence indicates that the incidence of neonatal encephalopathy caused by intrapartum hypoxia, in the absence of other abnormalities or antepartum events, is 1.6 per 10,000 cases [3]. They proposed the following criteria to identify an intrapartum event sufficient to lead to cerebral palsy:

1. Evidence of a metabolic acidosis in fetal umbilical cord arterial blood obtained at delivery (pH<7 and base deficit ≥12 mmol/L).
2. Early onset of severe or moderate neonatal encephalopathy in infants born at 34 or more weeks of gestation.
3. Cerebral palsy of the spastic quadriplegic or dyskinetic type.
4. Exclusion of other identifiable etiologies such as trauma, coagulation disorders, infectious conditions, or genetic disorders.

They also proposed criteria that suggest timing of the insult within 2 days of labor and delivery:

1. A sentinel (signal) hypoxic event occurring immediately before or during labor.
2. A sudden and sustained fetal bradycardia or the absence of fetal heart rate variability in the presence of persistent, late, or variable decelerations, usually after a hypoxic sentinel event when the pattern was previously normal.
3. Apgar scores of 0–3 beyond 5 min.
4. Onset of multi-system involvement within 72 h of birth.
5. Early imaging study showing evidence of acute nonfocal cerebral abnormality.

Prior to the publication of the above monograph, the task force sent a survey to ACOG fellows, which showed good clinical performance but weak knowledge of neonatal encephalopathy and cerebral palsy [6]. A survey of practicing obstetricians demonstrated improved knowledge of neonatal encephalopathy among those more familiar with this ACOG monograph [7]. However, more than one-third of respondents stated their knowledge was poor or deficient, and more than three-quarters thought their residency training in the subject was insufficient. These criteria have also been challenged as being too restrictive [8]. Other intrapartum factors may act alone or in concert with hypoxia to cause perinatal brain injury.

Pathophysiology

Perinatal brain injury has many possible etiologies, but the type most concerning in the intrapartum setting is hypoxic-ischemic injury. Decreased placental oxygen transfer causes fetal hypoxemia, which, if prolonged, causes tissue hypoxia. In a hypoxic environment, the Kreb's cycle of aerobic metabolism cannot be maintained and anaerobic metabolism prevails. During anaerobic metabolism, lactic acid is produced, causing a metabolic acidosis. This combination of metabolic acidosis and hypoxia is asphyxia. Hypoxic injury may be sufficient to cause necrosis or later apoptosis of neural tissue.

The fetal circulation initially responds to decreased oxygen by decreasing blood flow to certain organs (lungs, liver, kidney) in order to preserve flow of oxygenated blood to more critical organs (brain, heart, adrenal glands). As asphyxia continues, cerebral autoregulation fails, resulting in pressure-passive brain perfusion. With further insult, increasing acidosis causes myocardial suppression, which decreases cardiac output and blood pressure, causing further decreased brain perfusion and subsequent injury.

The timing and severity of the insult and gestational age at which the insult occurs are important factors that determine the location and extent of brain injury [9]. Injuries at different gestational ages can be categorized by the anatomic region affected: parasagittal, basal ganglia, periventricular white matter, focal/multifocal, or selective neuronal necrosis [3]. The developing brain of the *preterm infant* may be particularly susceptible to hypoxic injury during pregnancy [10]. Hypoxia may lead to periventricular leukomalacia and neuronal or axonal injury of the cerebral white matter, thalamus, basal ganglia, brainstem, and cerebellum [11]. Up to 50% of infants born at very low birth weights (<1,500 g) have imaging evidence of such injury [11]. The neuropathology relevant to the injury of the premature infant brain recently has been well described [11].

In the *term infant*, hypoxic injury typically affects gray matter. Among term infants born following a sentinel hypoxic event, the predominant areas of injury seen on imaging studies are the basal ganglia and thalamus [12]. Decreased cerebral perfusion also frequently occurs in watershed areas of cerebral circulation, leading to ischemic necrosis of vulnerable areas. Injury of the parasagittal area is frequently seen in term infants and is associated with the spastic quadriplegic form of cerebral palsy [3]. Evidence of white matter injury may also be seen but usually in association with central gray matter injury.

Intrapartum Complications

There are many intrapartum complications that have been associated with neonatal encephalopathy and cerebral palsy. Significant, or *sentinel*, events, such as placental abruption or other causes of hemorrhage, uterine rupture, and umbilical cord prolapse, may cause an abrupt decrease or loss of flow of oxygenated blood to the fetus [12]. Obstetric complications, including preterm labor and delivery, infection, preeclampsia, fetal malpresentation, and shoulder dystocia, may contribute to the development of neonatal encephalopathy and cerebral palsy.

Placental Abruption

Sentinel events during labor may cause significant reduction or complete disruption of oxygen transfer to the fetus. Placental abruption is the premature, acute or chronic, partial or complete, separation of the placenta from the uterus. It is associated with many etiologies, including trauma, hypertension, rapid uterine decompression, cocaine use, cigarette smoking, and maternal thrombophilias. The diagnosis of placental abruption is primarily clinical and should be considered in any pregnant patient presenting with vaginal bleeding, abdominal pain, and uterine contractions. Should placental abruption be serious, there may be a significant decrease in maternal–fetal gas exchange causing acute or chronic hypoxia in the fetus. Placental abruption may therefore be associated with perinatal brain injury [13]. Cerebral palsy has been reported in some cases of maternal trauma without abdominal injury [14]. These cases were likely due to subclinical abruption secondary to the trauma that lead to hypoxic brain injury.

Uterine Rupture

Uterine rupture is an uncommon complication during pregnancy. The most common type of uterine rupture is the separation of the uterine scar from a prior cesarean delivery. Uterine rupture can occur before or during labor and is the most concerning possible adverse outcome for women undergoing a trial of labor after a prior cesarean delivery. Rates of uterine rupture range from a low of 0.16% for women having a repeat cesarean before labor onset to 0.52% for women entering labor spontaneously and 0.77% for those with induced labor [15]. The rate of rupture with labor induced with prostaglandins is 2.4%, leading to the recommendation that these medications not be used for labor induction of a scarred uterus [15]. Uterine rupture can also occur, albeit much less frequently, with an unscarred uterus in the context of trauma, uterine anomalies, prior uterine surgery, induction of labor, fetal malpresentation, and labor dystocia. The rate of uterine rupture in an unscarred uterus is 1 in 17,000–20,000 [16].

Uterine rupture can be asymptomatic and not even detected if a vaginal delivery occurs. It may be asymptomatic and only discovered at the time of a repeat cesarean delivery. However, rupture of the uterus can be a true obstetric emergency, putting the life and health of both mother and fetus in grave danger. The mother is at risk for hemorrhage, necessitating an emergent laparotomy, possibly with hysterectomy in 0.2–3% of cases [17]. Should all or part of the fetus or placenta exit the uterus due to the rupture, the fetus is at risk for significant blood loss with resultant hypoxia. In a large study comparing a trial of labor with repeat cesarean delivery, the incidence of symptomatic uterine rupture was 0.7% in the nearly 18,000 women attempting a vaginal delivery after a prior cesarean [17]. In those cases, the incidence of hypoxic-ischemic encephalopathy was 6.2%. There were no cases of encephalopathy in the nearly 16,000 women who elected for repeat cesarean delivery.

Another study compared the neurological outcome of infants delivered after uterine rupture with those delivered after nonreassuring fetal monitoring and found a significantly increased risk for neonatal encephalopathy in the uterine rupture group (33% vs. 5%) [16]. The severity of the neurological injury was also greater in the group with uterine rupture. The risk of uterine rupture is low for pregnant women with a scarred uterus, but given the significant risk for perinatal brain injury if rupture occurs, women should be properly counseled about their options and only undergo a trial of labor if they are at low risk of rupture and have a reasonable chance at having a successful vaginal birth.

Umbilical Cord Prolapse

Another sentinel event that puts the fetus at risk for sudden hypoxia is prolapse of the umbilical cord, which occurs in less than 1% of pregnancies. Should the umbilical cord prolapse through the cervix prior to delivery, it may be compressed by the fetus, diminishing blood flow to the fetus and causing hypoxia. Historically, perinatal mortality due to umbilical cord prolapse approached 50% but with the increased number of cesarean deliveries and improved neonatal resuscitation, mortality is now less than 10% [18]. Cord prolapse typically occurs after spontaneous or artificial rupture of membranes, in situations where the fetus is not fully engaged in the pelvis. This can be due to a contracted pelvis, fetal malpresentation, or excessive manipulation of the fetus, among other causes. Once umbilical cord prolapse is diagnosed, the fetal presenting part should be elevated manually per the vagina and an emergent cesarean delivery accomplished. The time from diagnosis to delivery reduces but does not eliminate the risk of perinatal morbidity and mortality. Delivery after umbilical cord prolapse is associated with lower Apgar scores [19]. Even with delivery in a timely fashion, neonatal encephalopathy can occur [18].

The sentinel events discussed above are true obstetric emergencies, and once the diagnosis is made, the normal management is to proceed with an emergent cesarean delivery. Given that these events expose the fetus to a significant decrease or complete absence of oxygen, potentially accompanied by blood loss and decreased blood flow, one can anticipate that the longer the condition is present before delivery, the worse the perinatal outcome. ACOG recommends that a cesarean delivery should commence within 30 min of making the decision to operate [20]. This recommendation is arbitrary, rather than evidence-based, a fact acknowledged in the ACOG guidelines. There is certainly a need to deliver sooner in certain emergencies, but a delay of more than 30 min is not harmful in many circumstances. Intrapartum electronic fetal heart rate monitoring lacks precision in identifying the fetus developing metabolic acidosis, and most neonates delivered urgently for nonreassuring fetal status, even when born after 30 min, have normal cord gases [21].

In a large study with over 3,000 cesarean deliveries for emergency indications, 65% of the procedures started within 30 min of the decision to operate [22]. However, most of these cases were for nonreassuring fetal monitoring; of the procedures that were performed for a sentinel event (placental abruption, cord prolapse, uterine rupture), 98% were initiated within 30 min. The infants who were delivered within 30 min were actually more likely to be acidemic and require intubation in the operating room. This is likely secondary to the true emergent nature of their condition vs. the less specific diagnosis of a nonreassuring heart rate pattern. Of the infants delivered for an emergent indication beyond 30 min, 5% showed signs of compromise. Determining an ideal interval from decision to delivery is problematic since some infants delivered rapidly will experience compromise and some of those with a poor outcome for whom delivery was delayed may not have done better with a more rapid delivery [22].

Preeclampsia

Preeclampsia and other hypertensive disorders are very common, occurring in at least 5% of pregnancies [23]. Preeclampsia is a syndrome of hypertension and proteinuria during pregnancy. Other symptoms may also be present, including edema, headaches, visual disturbances, and epigastric pain [23]. The only treatment is delivery. Most commonly, the cases are mild, and maternal and neonatal outcomes are good. Some cases, however, can be quite severe, putting the mother and fetus at risk for long-term morbidity or mortality. The fetus is at risk for complications of prematurity, because preterm delivery is often the required treatment for severe cases. Preeclampsia has also been associated with neonatal encephalopathy and cerebral palsy in term neonates [24]. The causes of preeclampsia are not known but inflammation is thought to be a major factor. Inflammation may also play a role in the development of neonatal encephalopathy in the context of preeclampsia. Neonatal encephalopathy has been associated with preeclampsia independent of obstetric interventions and fetal acidemia [25]. It may be that the inflammatory process associated with preeclampsia causes direct fetal brain injury, and that this injury is exacerbated by intrapartum hypoxia.

Fever and Chorioamnionitis

Intrapartum maternal fever and chorioamnionitis, both inflammatory processes, have been linked to neonatal encephalopathy and cerebral palsy. Chorioamnionitis is diagnosed clinically when maternal fever and uterine tenderness are present. Maternal and/or fetal tachycardia is also often noted. The effects of chorioamnionitis on the fetus (e.g., low Apgar scores) can be similar to those of hypoxic injury, thereby confounding the determination of an injury. In a study population of nearly 240,000 infants with 109 cases of moderate to severe cerebral palsy not related to a developmental abnormality, there was a fourfold increased risk of cerebral palsy in the context of clinical chorioamnionitis [26]. In another study, neonates exposed to an intrapartum fever had a 3.1-fold increased risk of encephalopathy (8.3% vs. 2.7%), and those exposed to chorioamnionitis had a 5.4-fold increased risk (6.3% vs. 1.3%) [27]. The risk of neonatal encephalopathy was higher for infants delivered by ill women (those meeting criteria for chorioamnionitis) compared with those with fever alone, suggesting that the fetal response may vary based on the severity of maternal disease. Infants born to a febrile mother have been shown to be at higher risk of encephalopathy if they are acidotic (1.9% with maternal fever vs. 2.8% with acidosis vs. 12.5% with both) [28].

Many pathways have been proposed to explain the connection between chorioamnionitis and perinatal brain injury. Maternal fever, fetal cytokines, and/or fetal infection may lead to direct injury; inflammation of the placental membranes may adversely affect blood flow and gas exchange leading to hypoxic brain injury [26]. Ongoing research in this area may further elucidate the pathophysiology of this process, leading to treatments that could reduce fetal injury.

Shoulder Dystocia

Shoulder dystocia is an obstetric emergency that occurs when the fetal shoulder impacts behind the pubic symphysis after delivery of the head, preventing the remainder of the body from delivering. In a true shoulder dystocia, additional maneuvers beyond gentle downward traction on the fetal head are required to deliver the shoulder. Commonly, the obstetrician places a hand in the vagina to rotate the fetus obliquely to dislodge the shoulder and narrow the bisacromial distance. Delivery of the posterior arm also narrows the shoulder diameter and allows for delivery. The concerns for the fetus during shoulder dystocia are injury to the brachial plexus, causing nerve palsies, and hypoxia since the umbilical cord may be under tension and compressed. In one study, neonatal acidosis was diagnosed in 7% of deliveries complicated by shoulder dystocia with 1.5% of infants having neurological sequelae [29]. The risk for hypoxic injury because of a prolonged interval between delivery of the head and body of the infant must be balanced against the risk of brachial plexus injury due to the use of hastily performed or improper techniques to relieve the shoulder dystocia.

Breech Delivery

Vaginal breech delivery is another obstetric complication that may increase the risk for neonatal encephalopathy. Since the fetal head is disproportionately larger than the body, there is a possibility that during a breech delivery, there will be adequate space for delivery of the body, but the head may not have sufficient room to follow. Once the body has delivered, the umbilical cord is compressed between the fetal head and neck and cervix as it travels from the fetal abdomen to the placenta. If the head does not deliver in a timely fashion then there is risk for hypoxia. Management options for breech presentation include planned vaginal breech delivery, planned cesarean, and external cephalic version.

A multinational study, known as the Term Breech Trial, randomized over 2,000 women with breech presentation to planned vaginal delivery vs. planned cesarean delivery at term to determine which resulted in the best outcomes [30]. This study concluded that perinatal morbidity and mortality were significantly lower for patients having a cesarean delivery. Markers for perinatal asphyxia, the Apgar scores and umbilical artery pH with base deficit, were all significantly worse for breech infants born vaginally. Largely because of this study, singleton vaginal breech deliveries are now rarely performed in a nonemergent setting except by experienced operators with carefully selected patients. However, later analysis of this data showed that planned cesarean delivery is not associated with a reduction in the risk of death or developmental delay in children at 2 years of age [31]. In a review of over 32,000 deliveries in Scotland from 1985 to 2004, the rate of breech vaginal delivery decreased from 20% to 5% [32]. During the study period, the risk of perinatal death due to anoxia or mechanical causes decreased by 90% with planned cesarean delivery accounting for 16% of the reduction.

External cephalic version for breech presentation is performed by manual rotation of the fetus in a somersault fashion with the goal of converting to a cephalic presentation. The procedure is typically performed at 36–37 weeks of gestation with an average success rate of 58% [33]. The procedure is not without risk, however, with serious complications occurring in 0.24% of cases [34]. Some adverse outcomes for this method, including placental abruption, umbilical cord prolapse, and fetal bradycardia, may necessitate emergent cesarean delivery. Despite these risks, the overall outcome for breech fetuses is likely better with a protocol of planned cesarean delivery or external cephalic version than with planned vaginal breech delivery.

Intrapartum Identification of Asphyxia

Recent research suggests that there is a time interval between hypoxic insult and cell death that may be an opportunity for intervention [35]. The difficultly is to detect hypoxia when it occurs in order to allow sufficient time to act and prevent injury. In this section, we will review various methods that may be used to detect intrapartum hypoxia.

Cardiotocography, also known as electronic fetal monitoring, is the most commonly used intrapartum method to evaluate fetal status. Fetal heart rate and uterine contractions are monitored in a continuous fashion. The pattern is evaluated for heart rate, variability, and accelerations or decelerations of the rate. For purposes of standardizing communication about fetal heart rate patterns, the National Institute of Child Health and Human Development, ACOG, and the Society for Maternal-Fetal Medicine recently published revised definition and interpretation guidelines [36]. The authors recommend a three-tier system for classifying fetal heart rate patterns. Category I tracings have a normal heart rate, variability, and no concerning decelerations, and are associated with a fetus under no distress. Category III tracings show definitely worrisome findings in the heart rate, variability, and/or deceleration pattern and require further evaluation or intervention for fetal distress. Category II tracings show some reassuring and some nonreassuring features and need more evaluation to determine fetal status.

In the 1960s, cardiotocography was introduced into clinical practice based on the concept that it would improve neonatal neurologic outcomes. A recent Cochrane Review on the topic refuted this presumption [37]. The only improvement associated with cardiotocography was a 50% decrease in the incidence of neonatal seizures; however, this was not associated with any improvement in long-term outcome. There was no difference in the rates of cerebral palsy, infant mortality, or other neonatal outcome measures. Monitoring increases the rate of cesarean and operative vaginal deliveries, both of which increase morbidity to the mother. If electronic fetal monitoring is considered a screening test for a compromised fetus, it has a poor positive predictive value. Cardiotocography is most useful when it is normal. The likelihood of a fetus being compromised in the context of normal monitoring is very low. Fetal heart rate patterns can also be altered by many medications commonly used during labor and delivery, including magnesium sulfate, corticosteroids, and narcotic analgesics, further complicating the task of correctly identifying a hypoxic fetus [38].

The limitations of cardiotocography discussed above do not include the fact that interpretation of fetal heart rate monitoring is subject to significant inter and intraobserver variation [38, 39]. Various methods have been developed to minimize the effect of this limitation, including computerized analysis of the standard cardiotocography fetal heart rate pattern and the analysis of the actual fetal electrocardiogram (ECG). Computerized fetal cardiotocography uses software to analyze the continuous fetal heart rate for patterns related to hypoxia. There is no evidence that computerized analysis better predicts hypoxia when compared with standard visual analysis [40].

The fetal ECG can be obtained using the standard fetal scalp electrode that is occasionally used during labor. Systems have been developed to monitor the ST segment of the fetal ECG for changes that may be caused by hypoxia. A review of several studies using ST segment analysis in conjunction with standard cardiotocography vs. cardiotocography alone showed a reduction in infants with neonatal encephalopathy, fetal scalp pH samples during labor, and operative vaginal delivery [41]. ST segment analysis did not lead to a reduction in the number of infants with severe acidemia, cesarean delivery, Apgar score <7 at 5 min, or admission to a special care nursery. These technologies are being used by many labor and delivery suites but have not yet become standard.

Fetal pulse oximetry was developed to directly assess fetal oxygenation during labor, especially in the context of nonreassuring fetal heart rate monitoring. It is intended as an adjunct to and not a replacement for cardiotocography. During labor, after the fetal membranes have been ruptured, the monitor is passed through the cervix and adjusted to rest against the fetal back, cheek, or other part of the head. The monitor uses the standard technique of determining oxygen saturation by measuring the differential light absorption of oxygenated and deoxygenated blood.

Since the goal is to avoid fetal hypoxia during labor, this technique should be advantageous because it directly measures the oxygen saturation of the fetal blood. Animal and human studies have determined that a saturation of 30% or more is normal in the fetal environment [42]. Thus, in the context of nonreassuring cardiotocography, the fetal oxygen saturation could help discern which fetuses are actually hypoxic and when interventions should be performed to improve fetal status. Another benefit would be to allow labor to continue in the presence of nonreassuring fetal monitoring but normal pulse oximetry. Fetal pulse oximetry has not been shown to reduce the overall cesarean delivery rate, but may decrease the rate of cesarean delivery for nonreassuring fetal status and the incidence of intrapartum fetal scalp blood sampling [42, 43]. However, ACOG currently does not recommend the use of fetal pulse oximetry in clinical practice because of concerns of uncertain benefit and potentially falsely reassuring results [38].

Fetal scalp blood sampling has been examined extensively as a means to identify a hypoxic fetus during labor. Scalp pH and lactate levels have been evaluated as a means to identify potentially hypoxic fetuses. A pH of <7.2 and/or an elevated lactate of >4.8 mmol/L have been shown to correlate with fetal acidemia [44]. The two methods (pH and lactate) have been compared with each other and have similar predictive value for acidemia [45]. However, scalp pH determination requires seven times more fetal blood than lactate determination (35 vs. 5 μL). Also, lactate measurements are successful more often because of technical difficulties with measuring pH. Lactate levels may better predict neonatal encephalopathy than scalp pH alone [44]. The actual sampling of fetal scalp blood requires minimal technical skill, but does require special instruments and a cervix that has dilated 2–3 cm. However, having a lab continuously available to perform pH analysis is difficult to maintain. Many institutions no longer perform the procedure and report no increase in cesarean deliveries for fetal indications since discontinuation [46].

Another method of determining fetal status during labor is evaluation of the fetal response to stimulation. Stimulation can be direct (scalp stimulation) or indirect (vibroacoustic stimulation). Scalp stimulation is the gentle stroking of the fetal head for 15 s through a dilated cervix. Vibroacoustic stimulation uses a small device placed on the maternal abdomen that emits a 3–5 s auditory signal. A normal fetal response to either form of stimulation is a fetal heart rate acceleration of at least 15 beats per minute lasting for at least 15 s. Excessive stimulation should be avoided as it can cause a vagally mediated fetal heart rate deceleration. A metaanalysis of studies comparing stimulatory methods to direct fetal scalp pH sampling demonstrated that these methods are effective in ruling out significant fetal acidemia when there is a normal fetal response [47]. Because of the ease of performance and predictive ability, fetal stimulation has largely supplanted fetal scalp pH as a test to aid obstetricians in the evaluation of fetuses with abnormal fetal heart rate tracings during labor.

Meconium-stained amniotic fluid is commonly considered a sign of fetal distress. Intrauterine passage of meconium is likely part of the vagal response to stress, such as cord compression; however, it has been attributed to other causes, including post-term pregnancy and may occur without any apparent etiology. The primary concern with meconium-stained amniotic fluid is the development of meconium aspiration syndrome, a serious respiratory syndrome in the newborn, caused by pulmonary irritation from inhaled meconium. Meconium aspiration syndrome occurs in approximately 5% of infants born in the context of meconium-stained amniotic fluid [48]. It is associated with adverse pulmonary and neurological outcomes. Hypoxia before or during labor may trigger meconium passage and subsequent aspiration. In the postpartum setting, airways obstructed by plugs of meconium and/or pneumonitis caused by the inhaled meconium may impair oxygenation, worsening hypoxic injury. In one study, 21% of affected infants had developmental delay or cerebral palsy [49].

Various techniques have been used to reduce the incidence of meconium aspiration syndrome. Amnioinfusion, or the instillation of fluid into the uterine cavity via a transcervical catheter, can be used to reduce the umbilical cord compression and the resulting fetal heart rate decelerations. This technique has been used in the presence of meconium as well, both to reduce fetal gasping and inhalation of meconium and to dilute the potential toxic effect of the meconium on the fetal lungs. Unfortunately, this has not proven to reduce the risk of meconium aspiration syndrome [50]. Another practice used in the context of meconium-stained amniotic fluid was the suctioning of the oropharynx and nasopharynx of the infant following delivery of the head but before delivery of the shoulders. This practice is no longer recommended because of the lack of efficacy [51]. If the neonate is depressed following delivery, the trachea should be intubated, and any material below the glottis should be suctioned. The likely reason these interventions are ineffective is that the fetus has already aspirated meconium before the onset of labor. Thus, although intrapartum events may play a role, it is now believed that many – if not most – cases of meconium aspiration occur prior to labor and cannot be prevented [48].

Neonatal Markers of Hypoxic Injury

The Apgar score was first described in 1953 with the goal of providing a "basis for discussion and comparison of the results of obstetric practices, types of maternal pain relief, and the effects of resuscitation" [52]. Apgar scores are among the most commonly used perinatal outcome measures in research and are perceived by clinicians and patients as immediate indicators of perinatal outcome, both good and poor. Low Apgar scores have been associated with development of cerebral palsy. A population-based study of over 230,000 births in Norway demonstrated an 81-fold increased risk of cerebral palsy with Apgar scores of 3 or less at 1 min compared with a score of 7 or greater. If both the 1 and 5 min score was 3 or less, the risk increased 145-fold [53]. Using the Apgar score as the sole criteria to identify hypoxic injury is problematic. Many other factors can influence their interpretation, including gestational age, maternal medications, anesthesia and analgesia, neonatal disease and/or malformations, person assigning the score, and the nature of the neonatal resuscitation.

Umbilical artery pH is an objective method of determining fetal metabolic status at the time of delivery. A sample of blood is collected from the umbilical artery in a heparinized syringe. The pH and blood gas assessment is stable for 60 min in a clamped segment of cord and is likewise stable for 60 min in a heparinized syringe. ACOG recommends that cord blood be sent for analysis if there was a serious complication with the delivery or if there is a problem with the neonate's condition that persists at or beyond 5 min of life [54]. Some institutions choose to collect and analyze cord gases at all deliveries. Significant acidemia with an umbilical artery base deficit of 12 mmol/L or greater has been associated with an increased risk of neonatal encephalopathy as well as respiratory and other neonatal complications. In one study, the risk of moderate or severe neonatal complications was 10% in infants with a base deficit of 12–16 mmol/L and increased to 40% with a base deficit of >16 mmol/L [55]. As mentioned earlier, ACOG states that an umbilical artery pH <7.0 and a base deficit of 12 mmol/L or greater supports the diagnosis of an intrapartum hypoxic event sufficient to cause neonatal encephalopathy.

Other umbilical cord blood parameters beyond evaluation of the acid–base status have been shown to predict neonatal encephalopathy. Such measurements include umbilical cord lactate and nucleated red blood cell count. The umbilical cord lactate level, like scalp lactate, is a marker of acidemia, along with pH and base deficit. Although a general reference quotes an elevated lactate as >2.5 mmol/L [56], in one study the initial lactate level was significantly higher (>7.5 mmol/L or 67 mg/dL) in infants who developed moderate or severe encephalopathy [57], and this value had greater sensitivity and predictive value than cord pH or base deficit.

Nucleated red blood cells are found in the cord blood of healthy neonates, but their number has been shown to increase in the setting of both acute and chronic hypoxia. In response to hypoxia, fetal hematopoiesis accelerates to increase the oxygen-carrying capacity of the blood. As a result, the number of immature nucleated red blood cells in the circulation increases. It has been demonstrated that an elevated level of nucleated red blood cells in the newborn correlates with other markers of perinatal asphyxia and is predictive of neonatal encephalopathy [58]. In one study, an elevated nucleated red blood cell count in combination with elevated cord lactate have a combined sensitivity and specificity of 94% and 96%, respectively, in predicting hypoxic-ischemic encephalopathy [59]. Controversy exists regarding the timing and duration of hypoxic insult that leads to an elevation in the nucleated red blood cell count. There is evidence for both chronic hypoxia and acute intrapartum events playing a role [60]. Both acute and chronic hypoxia-ischemia can increase these counts, which limits their ability to determine the timing of the inciting injury [61].

Recent research has explored other neonatal biomarkers for perinatal brain injury. A metaanalysis demonstrated four markers potentially predictive of abnormal neurological outcomes: serum and cerebrospinal fluid interleukin-1b, serum interleukin-6, and cerebrospinal fluid neuron-specific enolase [62]. Further research is required to confirm these findings and develop therapies.

Interventions to Reduce Perinatal Brain Injury

In the preceding section, we evaluated several means of monitoring and evaluating a fetus during labor for the development of hypoxia. If a fetus was suspected to be hypoxic then delivery could be expedited, by a cesarean delivery if necessary, with the goal of preventing hypoxic brain injury. Other approaches to preventing perinatal brain injury due to intrapartum events are to completely avoid labor or to use some form of prophylaxis to decrease the risk of injury during labor.

Cesarean delivery before labor is most often performed for patients with a history of prior cesarean delivery who are not candidates for or do not desire a trial of labor. Increasingly, however, patients are requesting a cesarean delivery for other reasons, including fear of labor, convenience of scheduling, and pelvic floor preservation. There is insufficient evidence to support routinely offering this practice to patients especially in the context of increased risks of cesarean vs. vaginal delivery [63]. However, ACOG has deemed that it is ethical to perform the procedure for a requesting patient after well-informed consent [64]. Current evidence suggests that most cases of cerebral palsy are not due to intrapartum hypoxia and thus would not be prevented by cesarean delivery before labor. However, there are studies that show a decrease in neonatal encephalopathy among patients undergoing cesarean delivery before labor. In an Australian case-controlled study, there was an 83% reduction in risk of moderate to severe neonatal encephalopathy [24]. Authors of a review of the impact of cesarean delivery on neonatal morbidity extrapolated from this that the annual incidence of neonatal encephalopathy could be reduced from 11,400 to 1,900 if the at-risk population was delivered by cesarean [65]. Thus, there is evidence that cesarean delivery before labor may reduce the risk of neonatal encephalopathy; however, given the low incidence of neonatal injury, a large number of women would need to undergo cesarean delivery to prevent one case of neonatal encephalopathy.

Parenteral magnesium sulfate is a commonly used medication in obstetric practice. It is used to prevent seizures in the peripartum period in women with preeclampsia. It is also one of the several medications available for the inhibition of preterm labor. There has been much debate about the use of such medications, known as tocolytics, as many studies have not proven their effectiveness. A review of magnesium sulfate for the inhibition of preterm labor concluded that it is ineffective at delaying preterm delivery and is associated with increased infant mortality [66]. Although the use of magnesium sulfate as a tocolytic is not supported by current literature, there is increasing evidence for its potential as a neuroprotective agent for the fetus. Early retrospective studies suggested a decrease in cerebral palsy in preterm infants exposed to magnesium sulfate prior to delivery.

A recent randomized, double-blind, placebo-controlled trial was enrolled with over 2,200 patients to evaluate whether magnesium sulfate administered during preterm labor (24–31 weeks of gestation) would reduce a composite outcome of stillbirth, infant death, or cerebral palsy [67]. The study failed to find a reduction in this composite outcome, but among surviving infants, the rate of cerebral palsy was significantly reduced by nearly 50% (1.9% vs. 3.5%). A systematic review of studies of magnesium sulfate for fetal neuroprotection reached the same conclusion [68]. On the basis of this review, approximately 63 women would need to be treated with magnesium sulfate to prevent one case of cerebral palsy. The review also concluded there was no increased risk of infant mortality with the use of magnesium sulfate. Although these data are promising, some authorities recommend caution in adopting this therapy given the variation in the populations and protocols evaluated in the various studies [69]. More research is needed to elucidate the ideal selection of patients for magnesium sulfate therapy as well as the timing and method of administration.

In summary, the events of normal and complicated labor and delivery can put the fetus at risk for hypoxia with subsequent acidosis and injury. Current techniques are limited in their usefulness for determining at-risk fetuses during labor. This limits the ability of obstetricians to intervene by hastening delivery. However, it is also important to recognize that most of these injuries are not caused by peripartum events. Several markers have utility in identifying neonates that are at risk for injury. This is a critical step as neonatologists continue to develop early treatments for neonates suspected of being injured.

References

1. Little WJ (1862) On the influence of abnormal parturition, difficult labors, premature birth and asphyxia neonatorum on the mental and physical condition of the child, especially in relation to deformities. Trans Obstet Soc Lond 3:293–344.
2. Quilligan EJ, Paul RH (1975) Fetal monitoring: is it worth it? Obstet Gynecol 45:96–100.
3. American College of Obstetricians and Gynecologists (2003) Neonatal encephalopathy and cerebral palsy: defining the pathogenesis and pathophysiology. American College of Obstetricians & Gynecologists, Washington, DC.
4. Nelson KB (2002) The epidemiology of cerebral palsy in term infants. Ment Retard Dev Disabil Res Rev 8:146–150.
5. Graham EM, Ruis KA, Hartman AL, Northington FJ, Fox HE (2008) A systematic review of the role of intrapartum hypoxia-ischemia in the causation of neonatal encephalopathy. Am J Obstet Gynecol 199:587–595.
6. Hankins GD, Erickson K, Zinberg S, Schulkin J (2003) Neonatal encephalopathy and cerebral palsy: a knowledge survey of Fellows of The American College of Obstetricians and Gynecologists. Obstet Gynecol 101:11–17.

7. Morgan MA, Hankins GD, Zinberg S, Schulkin J (2005) Neonatal encephalopathy and cerebral palsy revisited: the current state of knowledge and the impact of American College of Obstetricians and Gynecologists task force report. J Perinatol 25:519–525.
8. Freeman RK (2008) Medical and legal implications for necessary requirements to diagnose damaging hypoxic-ischemic encephalopathy leading to later cerebral palsy. Am J Obstet Gynecol 199:585–586.
9. Scafidi J, Gallo V (2008) New concepts in perinatal hypoxia ischemia encephalopathy. Curr Neurol Neurosci Rep 8:130–138.
10. Billiards SS, Pierson CR, Haynes RL, Folkerth RD, Kinney HC (2006) Is the late preterm infant more vulnerable to gray matter injury than the term infant? Clin Perinatol 33:915–933, abstract x–xi.
11. Volpe JJ (2009) Brain injury in premature infants: a complex amalgam of destructive and developmental disturbances. Lancet Neurol 8:110–124.
12. Okereafor A, Allsop J, Counsell SJ et al (2008) Patterns of brain injury in neonates exposed to perinatal sentinel events. Pediatrics 121:906–914.
13. Kayani SI, Walkinshaw SA, Preston C (2003) Pregnancy outcome in severe placental abruption. BJOG 110:679–683.
14. Hayes B, Ryan S, Stephenson JB, King MD (2007) Cerebral palsy after maternal trauma in pregnancy. Dev Med Child Neurol 49:700–706.
15. American College of Obstetricians and Gynecologists Committee on Obstetric Practice (2006) ACOG Committee Opinion No. 342: induction of labor for vaginal birth after cesarean delivery. Obstet Gynecol 108:465–468.
16. Martínez-Biarge M, García-Alix A, García-Benasach F et al (2008) Neonatal neurological morbidity associated with uterine rupture. J Perinat Med 36:536–542.
17. Landon MB, Hauth JC, Leveno KJ et al (2004) Maternal and perinatal outcomes associated with a trial of labor after prior cesarean delivery. N Engl J Med 351:2581–2589.
18. Lin MG (2006) Umbilical cord prolapse. Obstet Gynecol Surv 61:269–277.
19. Dilbaz B, Ozturkoglu E, Dilbaz S, Ozturk N, Sivaslioglu AA, Haberal A (2006) Risk factors and perinatal outcomes associated with umbilical cord prolapse. Arch Gynecol Obstet 274:104–107.
20. American Academy of Pediatrics, American College of Obstetricians and Gynecologists (2007) Guidelines for perinatal care, 6th edn. ACOG, Washington, DC.
21. Holcroft CJ, Graham EM, Aina-Mumuney A, Rai KK, Henderson JL, Penning DH (2005) Cord gas analysis, decision-to-delivery interval, and the 30-minute rule for emergency cesareans. J Perinatol 25:229–235.
22. Bloom SL, Leveno KJ, Spong CY et al (2006) Decision-to-incision times and maternal and infant outcomes. Obstet Gynecol 108:6–11.
23. ACOG Committee on Practice Bulletins – Obstetrics (2002) ACOG practice bulletin. Diagnosis and management of preeclampsia and eclampsia. Number 33, January 2002. Obstet Gynecol 99:159–167.
24. Badawi N, Kurinczuk JJ, Keogh JM et al (1998) Antepartum risk factors for newborn encephalopathy: the Western Australian case-control study. BMJ 317:1549–1553.
25. Impey L, Greenwood C, Sheil O, MacQuillan K, Reynolds M, Redman C (2001) The relation between pre-eclampsia at term and neonatal encephalopathy. Arch Dis Child Fetal Neonatal Ed 85:F170–F172.
26. Wu YW, Escobar GJ, Grether JK, Croen LA, Greene JD, Newman TB (2003) Chorioamnionitis and cerebral palsy in term and near-term infants. JAMA 290:2677–2684.
27. Blume HK, Li CI, Loch CM, Koepsell TD (2008) Intrapartum fever and chorioamnionitis as risks for encephalopathy in term newborns: a case-control study. Dev Med Child Neurol 50:19–24.
28. Impey LW, Greenwood CE, Black RS, Yeh PS, Sheil O, Doyle P (2008) The relationship between intrapartum maternal fever and neonatal acidosis as risk factors for neonatal encephalopathy. Am J Obstet Gynecol 198:49.e1–49.e6.
29. MacKenzie IZ, Shah M, Lean K, Dutton S, Newdick H, Tucker DE (2007) Management of shoulder dystocia: trends in incidence and maternal and neonatal morbidity. Obstet Gynecol 110:1059–1068.

30. Hannah ME, Hannah WJ, Hewson SA, Hodnett ED, Saigal S, Willan AR (2000) Planned caesarean section versus planned vaginal birth for breech presentation at term: a randomised multicentre trial. Term Breech Trial Collaborative Group. Lancet 356:1375–1383.
31. Whyte H, Hannah ME, Saigal S et al (2004) Outcomes of children at 2 years after planned cesarean birth versus planned vaginal birth for breech presentation at term: the International Randomized Term Breech Trial. Am J Obstet Gynecol 191:864–871.
32. Pasupathy D, Wood AM, Pell JP, Fleming M, Smith GC (2009) Time trend in the risk of delivery-related perinatal and neonatal death associated with breech presentation at term. Int J Epidemiol 38(2):490–498.
33. Committee on Practice Guidelines–Gynecology, American College of Obstetrics and Gynecologists (2001) External cephalic version. Int J Gynaecol Obstet 72(2):198–204. PMID: 11291628 [PubMed-indexed for MEDLINE].
34. Grootscholten K, Kok M, Oei SG, Mol BW, van der Post JA (2008) External cephalic version-related risks: a meta-analysis. Obstet Gynecol 112:1143–1151.
35. du Plessis AJ, Volpe JJ (2002) Perinatal brain injury in the preterm and term newborn. Curr Opin Neurol 15:151.
36. Macones GA, Hankins GD, Spong CY, Hauth J, Moore T (2008) The 2008 National Institute of Child Health and Human Development workshop report on electronic fetal monitoring: update on definitions, interpretation, and research guidelines. Obstet Gynecol 112: 661–666.
37. Alfirevic Z, Devane D, Gyte (2006) Continuous cardiotocography (CTG) as a form of electronic fetal monitoring (EFM) for fetal assessment during labour. Cochrane Database Syst Rev (3):CD006066.
38. American College of Obstetricians and Gynecologists (2005) ACOG practice bulletin. Clinical management guidelines for obstetrician-gynecologists, Number 70, December 2005 (replaces practice bulletin Number 62, May 2005). Intrapartum fetal heart rate monitoring. Obstet Gynecol 106:1453–1460.
39. Chauhan SP, Klauser CK, Woodring TC, Sanderson M, Magann EF, Morrison JC (2008) Intrapartum nonreassuring fetal heart rate tracing and prediction of adverse outcomes: interobserver variability. Am J Obstet Gynecol 199:623.e1–623.e5.
40. Agrawal SK, Doucette F, Gratton R, Richardson B, Gagnon R (2003) Intrapartum computerized fetal heart rate parameters and metabolic acidosis at birth. Obstet Gynecol 102:731–738.
41. Neilson JP (2006) Fetal electrocardiogram (ECG) for fetal monitoring during labour. Cochrane Database Syst Rev (3):CD000116.
42. East CE, Chan FY, Colditz PB, Begg LM (2007) Fetal pulse oximetry for fetal assessment in labour. Cochrane Database Syst Rev (2):CD004075.
43. Kühnert M, Schmidt S (2004) Intrapartum management of nonreassuring fetal heart rate patterns: a randomized controlled trial of fetal pulse oximetry. Am J Obstet Gynecol 191:1989–1995.
44. Kruger K, Hallberg B, Blennow M, Kublickas M, Westgren M (1999) Predictive value of fetal scalp blood lactate concentration and pH as markers of neurologic disability. Am J Obstet Gynecol 181:1072–1078.
45. Wiberg-Itzel E, Lipponer C, Norman M et al (2008) Determination of pH or lactate in fetal scalp blood in management of intrapartum fetal distress: randomised controlled multicentre trial. BMJ 336:1284–1287.
46. Goodwin TM, Milner-Masterson L, Paul RH (1994) Elimination of fetal scalp blood sampling on a large clinical service. Obstet Gynecol 83:971–974.
47. Skupski DW, Rosenberg CR, Eglinton GS (2002) Intrapartum fetal stimulation tests: a meta-analysis. Obstet Gynecol 99:129–134.
48. Ross MG (2005) Meconium aspiration syndrome – more than intrapartum meconium. N Engl J Med 353:946–948.
49. Beligere N, Rao R (2008) Neurodevelopmental outcome of infants with meconium aspiration syndrome: report of a study and literature review. J Perinatol 28(Suppl 3):S93–S101.

50. Fraser WD, Hofmeyr J, Lede R et al (2005) Amnioinfusion for the prevention of the meconium aspiration syndrome. N Engl J Med 353:909–917.
51. Committee on Obstetric Practice, American College of Obstetricians and Gynecologists (2007) ACOG Committee Opinion No. 379: management of delivery of a newborn with meconium-stained amniotic fluid. Obstet Gynecol 110:739.
52. Apgar V (1953) A proposal for a new method of evaluation of the newborn infant. Curr Res Anesth Analg 32:260–267.
53. Moster D, Lie RT, Irgens LM, Bjerkedal T, Markestad T (2001) The association of Apgar score with subsequent death and cerebral palsy: a population-based study in term infants. J Pediatr 138:798–803.
54. American College of Obstetricians and Gynecologists (2006) Umbilical cord blood gas and acid-base analysis. ACOG Committe Opinion No. 348. Obstet Gynecol 108:1319–1322.
55. Low JA, Lindsay BG, Derrick EJ (1997) Threshold of metabolic acidosis associated with newborn complications. Am J Obstet Gynecol 177:1391–1394.
56. Gunn VL, Nechyba C (eds) (2002) Blood chemistries and body fluids. The Harriet Lane handbook. Mosby, Philadelphia, PA.
57. Shah S, Tracy M, Smyth J (2004) Postnatal lactate as an early predictor of short-term outcome after intrapartum asphyxia. J Perinatol 24:16–20.
58. Ghosh B, Mittal S, Kumar S, Dadhwal V (2003) Prediction of perinatal asphyxia with nucleated red blood cells in cord blood of newborns. Int J Gynaecol Obstet 81:267–271.
59. Haiju Z, Suyuan H, Xiufang F, Lu Y, Sun R (2008) The combined detection of umbilical cord nucleated red blood cells and lactate: early prediction of neonatal hypoxic ischemic encephalopathy. J Perinat Med 36:240–247.
60. Redline RW (2008) Elevated circulating fetal nucleated red blood cells and placental pathology in term infants who develop cerebral palsy. Hum Pathol 39:1378–1384.
61. Silva AM, Smith RN, Lehmann CU, Johnson EA, Holcroft CJ, Graham EM (2006) Neonatal nucleated red blood cells and the prediction of cerebral white matter injury in preterm infants. Obstet Gynecol 107:550–556.
62. Ramaswamy V, Horton J, Vandermeer B, Buscemi N, Miller S, Yager J (2009) Systematic review of biomarkers of brain injury in term neonatal encephalopathy. Pediatr Neurol 40:215–226.
63. NIH (2006) NIH State-of-the-Science Conference Statement on cesarean delivery on maternal request. NIH Consens State Sci Statements 23:1–29.
64. American College of Obstetricians and Gynecologists (2007) ACOG Committee Opinion No. 394, December 2007. Cesarean delivery on maternal request. Obstet Gynecol 110:1501.
65. Hankins GD, Clark SM, Munn MB (2006) Cesarean section on request at 39 weeks: impact on shoulder dystocia, fetal trauma, neonatal encephalopathy, and intrauterine fetal demise. Semin Perinatol 30:276–287.
66. Crowther CA, Hiller JE, Doyle LW (2002) Magnesium sulphate for preventing preterm birth in threatened preterm labour. Cochrane Database Syst Rev (4):CD001060.
67. Rouse DJ, Hirtz DG, Thom E et al (2008) A randomized, controlled trial of magnesium sulfate for the prevention of cerebral palsy. N Engl J Med 359:895–905.
68. Doyle LW, Crowther CA, Middleton P, Marret S, Rouse D (2009) Magnesium sulphate for women at risk of preterm birth for neuroprotection of the fetus. Cochrane Database Syst Rev (1):CD004661.
69. Stanley FJ, Crowther C (2008) Antenatal magnesium sulfate for neuroprotection before preterm birth? N Engl J Med 359:962–964.

30. Hagan MJ, Hodge V, [illegible] et al (2008) [illegible] asphyxia syndrome. N Engl J Med [illegible]
31. [illegible]
32. [illegible]
33. [illegible]
34. [illegible]
35. [illegible]
36. [illegible]
37. [illegible]
38. [illegible]

Chapter 7
Activation of the Maternal Immune System as a Risk Factor for Neuropsychiatric Disorders

Stephen E.P. Smith, Elaine Hsiao, and Paul H. Patterson

Keywords Schizophrenia • Autism • Maternal immune activation • Poly(I:C) • Influenza • Prenatal infection

Introduction

Surprisingly little is known about the subtle changes in the brain that result in neuropsychiatric disorders such as schizophrenia and autism. These disorders are defined by behavioral symptoms. A child presenting before the age of three with abnormal language acquisition and reciprocal social interactions as well as repetitive, stereotyped behaviors is diagnosed as autistic. A young adult who reports positive (hallucinations or delusions) and negative (depression, apathy, social withdrawal) symptoms is diagnosed as schizophrenic. Although these symptoms are dramatic, the brain looks grossly normal, and even upon microscopic examination, the neuropathology is subtle. However, studies of postmortem protein and gene expression have identified numerous changes, including many in molecules involved in immune regulation, synaptic function, and myelin. Unfortunately, our understanding of the normal function of the brain is not sufficiently developed to form a coherent model of the malfunction in these disorders.

Another approach to these disorders is to identify and study the risk factors that contribute to their development. Despite the postnatal onset of clinical symptoms, evidence suggests that both schizophrenia and autism have their origins in early brain development. Retrospective examination of school records and home videos

S.E.P. Smith (✉)
Biology Division, California Institute of Technology, Pasadena, CA 91125, USA
and
Departments of Neurology and Pathology, Harvard Medical School, Beth Israel Deaconess Medical Center, 330 Brookline Avenue, E/CLS-717, Boston, MA 02215, USA
e-mail: ssmith13@bidmc.harvard.edu

A.W. Zimmerman and S.L. Connors (eds.), *Maternal Influences on Fetal Neurodevelopment: Clinical and Research Aspects*,
DOI 10.1007/978-1-60327-921-5_7, © Springer Science+Business Media, LLC 2010

of children who subsequently display these disorders portray their early behaviors as "preschizophrenic" or "preautistic" [1]. Moreover, postmortem examination of the brains of at least some affected individuals reveals subtle changes in the number, distribution, or alignment of neurons, suggesting these abnormalities had their origins in utero [2]. Therefore, taking a bottom-up approach of identifying risk factors and tracing their mechanism of action may be a productive path toward understanding these complex disorders with heterogeneous symptoms.

Risk Factors Contributing to Schizophrenia and Autism

Although genetics clearly plays a major role in schizophrenia and autism, the complex nature of inheritance continues to puzzle researchers. Autism is often referred to as one of the most heritable neurodevelopmental disorders, with twin studies often cited as evidence: about 90% of identical twins are concordant for autism, while that number falls to roughly 10% for fraternal twins. Similar data exists for schizophrenia, with concordance rates of approximately 50% in monozygotic vs. close to 17% in dizygotic twins [3]. Despite the relatively high heritability, however, only about 5–10% of autism cases can be attributed to known chromosomal abnormalities or single gene mutations. Similarly, with notable exceptions, such as the DISC1 gene, very few cases of schizophrenia can be traced to a known genetic cause. Moreover, large, statistically powerful genetic association studies of patient populations have failed to identify more than a handful of implicated genetic variants, and these carry a small statistical effect size [4, 5].

Several theories, reviewed elsewhere, attempt to reconcile the high apparent heritability of neuropsychiatric disorders with the difficulty in identifying genetic risk factors. An early explanation is the "multiple interacting genes" hypothesis, which states that the inheritance involves multiple genes, each of which displays incomplete penetrance (e.g., [5]). A more recent theory posits that de novo genetic mutations and chromosomal rearrangements contribute to risk. Evidence showing an elevated rate of microdeletions and microduplications in autistic and schizophrenic patient samples as well as associations of autism with increasing parental age lends credence to this hypothesis [6]. Although de novo microdeletions and duplications would explain why so few candidate single gene mutations have been identified in studies of multiple populations, they do not explain the high heritability estimates of these disorders. Similarly, epigenetic regulation is postulated to be involved in autism and schizophrenia (for review see [7]).

Environmental factors may also contribute to the difficulties in genetic analyses. Proof of principle for a role of environmental factors is provided by studies showing that exposure to thalidomide [8], valproic acid [9], or prenatal infections [10–12] strongly increase the risk for autism and schizophrenia in the offspring. Therefore, it is possible that environmental risk factors contribute to genetic susceptibility or that environmental factors require a susceptibility genotype to increase the risk. It is important to understand all of the risk factors, genetic, epigenetic, and environmental, in order to develop a complete understanding of, and more effective treatments for, these disorders.

Maternal Infection: An Environmental Risk Factor

Maternal infection is among the most studied and best established nongenetic risk factor for schizophrenia. The connection between schizophrenia and maternal influenza infection was first indicated by the work of Mednick et al. [13]. Since then, over 25 studies have examined the rate of schizophrenia in populations, which were in utero during influenza epidemics, and the majority found an increased incidence among the exposed offspring (reviewed in [14]). However, such studies are population-based; they are unable to document a one-to-one relationship between respiratory infection in individual mothers and later development of schizophrenia in their children. To overcome this limitation, Brown, Susser, and colleagues examined a large pool of archived maternal serum samples that were collected in the 1960s and linked to detailed medical records of both mothers and children. They found that in cases where they were able to confirm maternal influenza infection by assaying the serum, the offspring were 3–7 times more likely than controls to develop schizophrenia [15]. Because of the high prevalence of influenza, they estimated that up to 21% of all schizophrenia cases may be traced to maternal influenza infection. A separate study examining Danish medical records found an eightfold increase in risk for schizophrenia following maternal influenza infection [16]. Further studies have found links between elevated levels of cord blood cytokines and development of schizophrenia, strengthening the association of influenza infection with schizophrenia risk [10, 17]. In addition, serological studies have found links between schizophrenia and maternal toxoplasmosis [18, 19], rubella [20], genital/reproductive [21], and bacterial [22] infections. Thus, many different pathogens increase the risk of neuropsychiatric disease in the offspring.

An important insight into the heterogeneity of the schizophrenia phenotype and its risk factors was revealed by further study of the population being monitored by Brown et al. Comparison of schizophrenics born to infected vs. noninfected mothers showed both anatomical (enlarged cavum septum pellucidum) and cognitive differences between the two groups [23, 24]. It will be interesting to see if these two groups also display divergent responses to antipsychotic drugs, and what the distribution of candidate gene variants for schizophrenia may be.

Similar, but considerably less extensive, epidemiological data exist for autism. Rubella epidemics in the 1960s were associated with a striking elevation in the incidence of autism in children born to infected mothers [12, 25]. Studies of other maternal infections, such as toxoplasmosis, syphilis, varicella, and rubeola, also support the idea that several types of maternal infections can be risk factors for autism (for a review, see [26]), although the number of individuals in these studies is very small.

Animal Models of Immune Activation

Maternal Influenza Infection

Both the epidemiologic data and the serologic studies show that infection during early-to-mid pregnancy results in the greatest schizophrenia risk in the offspring [15, 27]. To produce an animal model that most closely replicates the human data, Shi et al. [28] inoculated mice intranasally with a human H1N1 influenza virus on day 9.5 of pregnancy, which is approximately equivalent to early second trimester in humans. The pregnant dams develop lung pathology, show noticeable sickness behavior, and have elevated serum levels of several immune markers for approximately 7 days, covering the second half of the pregnancy. The adult offspring of influenza-infected mice appear superficially normal, but display several behavioral abnormalities that are highly relevant to schizophrenia and autism. For example, they have lower prepulse inhibition (PPI) than controls, and this deficit is rescued by acute treatment with antipsychotic drugs. PPI is abnormal in schizophrenia [29, 30] and autism [31], as well as in other neuropsychiatric disorders. The offspring of influenza-infected mice also display heightened anxiety in the open field, and they show a reduced number of social interactions when placed in an open field with an unfamiliar, same-gender animal of the same species [28]. These offspring also display several histochemical abnormalities that are reminiscent of those found in autism or schizophrenia, such as reduced reelin [32] and parvalbumin (PV) [33] immunoreactivity in the brain. They also display a spatially restricted deficit in Purkinje cells (PCs) in lobules VI and VII of the cerebellum [34], which is consistent with the most commonly reported histological abnormality in autism (reviewed in [35]).

Maternal Immune Activation

The fact that a wide variety of pathogens are able to increase schizophrenia risk in the offspring suggests that these pathogens alter fetal development by one or more common mechanisms. Direct infection of the fetus seems an obvious hypothesis. However, most of the implicated infections are confined to specific areas of the body, and do not involve the fetus. For example, influenza infection is confined to the lungs, since the membrane proteins that allow viral entry into a cell are found only in the lungs [36]. In the mouse model of maternal influenza infection, using a sensitive RT-PCR assay that can detect as little as one plaque-forming unit of virus, no virus is detected in the exposed fetuses [37]. Maternal immune activation is also a plausible mechanism; all of the implicated pathogens trigger a maternal immune response, which will alter the biochemical environment of the fetus. In strong support of this hypothesis, two rodent models have been developed in which behavioral deficits are induced in adult offspring by directly activating the maternal immune system in the absence of infectious organisms.

Maternal Poly(I:C) Administration

Poly(I:C) is a synthetic, double-stranded RNA that is a potent agonist of the toll-like receptor-3 (TLR-3). Toll-like receptors are sentinels of the innate immune system, which recognize markers of infection such as the bacterial flagellar protein flagellin or the bacterial cell wall component lipopolysaccharide (LPS). TLR-3 recognizes double-stranded ribonucleic acid (RNA), which is regarded by the innate immune system as an indication of viral infection. Hence, the activation of TLR-3 induces a proinflammatory cascade that results in the production of antiviral cytokines and chemokines. Injection of poly(I:C) in the pregnant rodent at midgestation produces offspring that are remarkably similar to the offspring of mice given a flu infection. These offspring display deficits in PPI, latent inhibition, open field exploration, and social interaction that are almost identical to those exhibited by offspring of maternal influenza infection [38]. Moreover, investigators using the poly(I:C) model of MIA have shown that the PPI and latent inhibition deficits occur in postpubertal but not in juvenile rats, mimicking the adult-onset nature of schizophrenia [39, 40]. These deficits are reversed by acute antipsychotic drug treatment. Maternal poly(I:C) also causes other changes in offspring that are relevant to autism or schizophrenia: A PC deficit [31], increased dopamine release and turnover [41], reduced PV expression [33], and increased $GABA_A$ receptor expression [42] in the adult offspring.

Maternal LPS Administration

MIA can also be induced by the injection of bacterial (LPS), a natural ligand for toll-like receptor-4 (TLR-4). Intrauterine bacterial infection is commonly associated with preterm birth, obstetrical complications, and cerebral palsy [43]. There is evidence linking maternal bacterial infection with schizophrenia in the offspring [44, 45], and ligands binding TLR-4 activate many of the same signaling pathways as TLR-3. The specific combination of cytokine and chemokine activation for TLR-4 is different than that for TLR-3, but like poly(I:C), LPS produces very strong, transient immune activation. Thus, bacterial and viral infections can induce similar responses. Indeed, some of the behavioral abnormalities seen in rodent offspring of poly(I:C)-treated mothers are also observed in the offspring of LPS-treated mothers. PPI deficits are often reported, as are increased anxiety, social interaction deficits, and abnormalities in dopamine-related behaviors [43, 46–49]. Histological findings in the LPS model include fewer, more densely packed neurons in the hippocampus, increased staining of microglia and astroglia, altered tyrosine hydroxylase staining (a marker for catecholamine neurotransmitters), and decreased myelin basic protein staining [46, 48, 50–52].

Other Protocols for Maternal Immune Activation

Bacteria commonly associated with periodontal disease have been isolated from preterm placentas, and epidemiological evidence links schizophrenia risk to both periodontal disease and preterm birth [53]. Mouse models involving injection of periodontal bacteria into a pregnant dam result in fetal growth restriction and mortality [53]. Injection of LPS directly into the fetus also results in low birth weight and fetal mortality (reviewed in [54]).

There is also a large body of research on maternal stress causing abnormal behavior in the adult offspring. Epidemiological evidence links maternal stress caused by diverse stimuli such as war or death of a loved one to schizophrenia in the offspring (reviewed in [55]), and animal models document behavioral changes in the offspring of rats exposed to daily restraint stress. Barbazanges et al. [56] showed that this effect of stress is mediated by glucocorticoids, which can also have strong modulatory effects on the innate immune system.

Mechanisms Underlying MIA-Induced Behavioral Abnormalities

Cytokines are small (8–30 kDa, in weight) signaling proteins that are released into the serum in response to a wide range of stimuli including infection, immune activation, and stress. Thus, even though overt signs of an influenza infection may be confined to the lungs, cytokines produced there will have access to the placenta and fetus. Cytokines signal through several important kinase pathways, including those of NF-*k*B, MAPK, and JAK/STAT. During brain development, these same pathways control functions such as neuronal differentiation and migration, axon pathfinding, and synapse formation. Moreover, many cytokine receptors are expressed in both developing and mature neurons and can cause morphological and functional changes in those cells [57–59]. Since many cytokines are able to cross from maternal circulation into the fetal brain, the resulting disruption in neuronal homeostasis could account for the long-term effects in the offspring.

Mounting evidence shows that the maternal inflammatory response does, in fact, cross the placenta and affect the fetus. Radiolabeling studies have demonstrated that the labeled cytokines interleukin (IL)-6 and IL-2 enter the fetus when injected into pregnant dams [60, 61]. Several groups have shown increased cytokine protein in the fetal brain following MIA induced by LPS or poly(I:C) injection into the mother during gestation. Increases in cytokine messenger RNA (mRNA) have been found in fetal brain as well, making it unclear if the cytokines found in the fetal brain are of maternal or fetal origin. Both protein and mRNA of IL-1β, IL-6, tumor necrosis factor α (TNF-α), and IL-10 are upregulated at various time points ranging

Table 7.1 Maternal immune activation elevates cytokines in fetal brain

Treatment	Findings	References
0.05 mg/kg LPS I.P. E18 rat	No change in TNFα, IL1β, IL6	[108]
0.12 mg/kg LPS I.P. E17 mouse	IL-6 increased	[109]
1 mg/kg LPS I.P. E18 rat	TNFα, IL-1β, iNOS increased[a]	[52]
50 μg LPS I.P. E18 mouse	IL-1β, IL-6, MCP-1, VEGF increased[a]	[110]
2.5 mg/kg LPS I.P. E16 rat	TNFα increased	[111]
4 mg/kg LPS I.P. E18 rat	TNFα, IL-1β increased[a]	[50]
20 mg/kg poly(I:C) I.P. E16 rat	No change in TNFα	[112]
2 mg/kg poly(I:C) I.V. E9 mouse	TNFα, IL-1β, IL-6, IL-10 increased	[62]
5 mg/kg poly(I:C) I.V. E9 mouse	IL-1β, IL-6 increased	[113]
5 mg/kg poly(I:C) I.V. E17 mouse	IL-1β, IL-6, IL-10 increased	[114]

Although some authors report no changes in cytokine levels, the majority of studies show significant increases. The studies that report no changes used less intense methods of immune activation (a lower dose of LPS or intraperitoneal (I.P.) instead of intraventricular (I.V.) administration of poly(I:C)). Assays are for cytokine protein, except where noted

[a]RNA assayed

from 2 to 24 h postinjection of LPS or poly(I:C) (see Table 7.1). This response seems to be dose-dependent, as groups that use more intense treatments (higher doses of poly(I:C) or LPS; intravenous vs. intraperitoneal injection) tend to see more robust increases in fetal brain cytokines. Recent work from our group has found evidence of induction of several response genes downstream from IL-6, confirming that cytokines in the fetal brain have significant biological effects (E. Hsiao and P.H. Patterson, unpublished).

Manipulation of cytokine expression during MIA has demonstrated that cytokine signaling plays a key role in the development of abnormal behavior. Meyer et al. were able to prevent the development of behavioral abnormalities in the offspring of MIA dams by overexpression of IL-10, an anti-inflammatory cytokine [62]. IL-10 can also protect against white matter damage caused by maternal *E. coli* infection [63]. Similarly, antiTNFα antibodies can reduce the growth restriction and fetal loss caused by LPS injection, and conversely, injection of TNFα can mimic the effects of LPS [64, 65].

Work from our laboratory has demonstrated that IL-6 plays a critical role in the effect of poly(I:C)-induced MIA [38]. We screened for cytokines that might mediate the effect of MIA by coinjecting poly(I:C) and various anticytokine antibodies into pregnant mice. We found that coinjection of antiIL-6 prevented the behavioral deficits caused by poly(I:C). Antibodies to IL-1β, TNFα, and interferon-γ had no significant effect. Moreover, blocking IL-6 also prevented maternal poly(I:C)-induced gene expression changes in the adult offspring brain, as measured by RNA microarray. In addition, IL-6 knockout mice did not show behavioral deficits after MIA. Conversely, a single injection of recombinant IL-6 (but not the other cytokines tested) into pregnant mice caused behavioral deficits in their offspring similar to MIA. Samuelsson et al. have reported long-term effects of multiple IL-6 injections in pregnant rats [66]. The offspring showed deficits in learning as well as elevated serum levels of IL-6 as adults, which may mimic the ongoing immune activation observed in neuropsychiatric illness (see below).

These animal models demonstrate that MIA, particularly maternal IL-6 activation, causes long-term changes in the behavior of the offspring that are reminiscent of symptoms seen in certain human neuropsychiatric illnesses (Fig. 7.1). But how could a transient increase in a single immune signaling molecule such as IL-6 (albeit at a critical time in development) precipitate such life-long changes? It is possible that normal development of homeostatic regulatory mechanisms is disrupted, leading to dysregulation of gene expression not only in the brain, but also in the immune system.

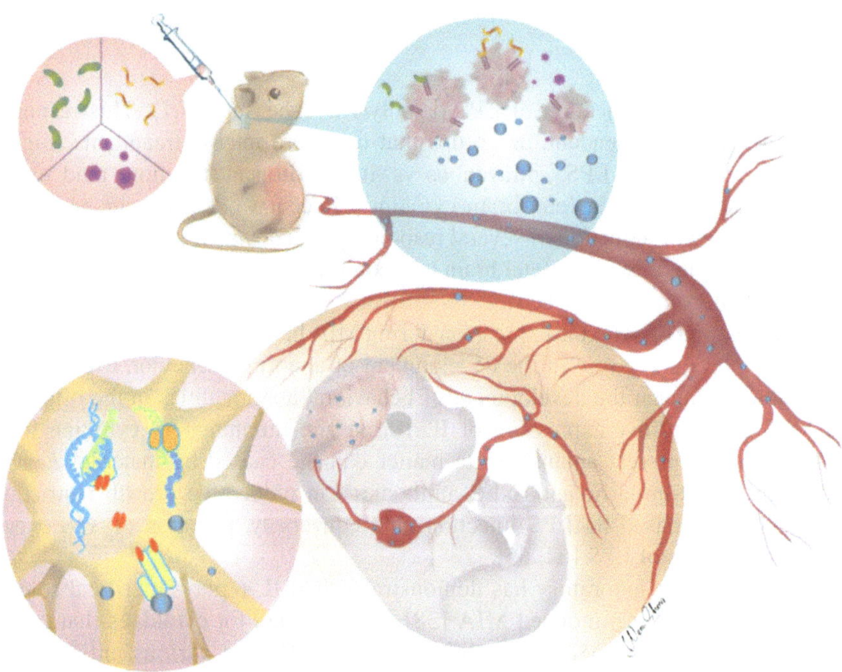

Fig. 7.1 Proposed mechanism through which maternal immune activation (MIA) leads to behavioral abnormalities. Clockwise from *top left*: A pregnant rodent is infected with influenza virus, or injected with LPS or poly(I:C). Each inflammatory stimulus induces the production of proinflammatory cytokines, including IL-6 (*blue spheres*) by maternal immune cells. These cytokines travel through the maternal circulation, cross the placenta, and enter the fetal circulation (see text). Proinflammatory cytokines, cytokine mRNA, and mRNA of cytokine-induced genes are found in the fetal brain following MIA. The alterations to fetal brain development caused by increased cytokine signaling likely contribute to the abnormal behavior in the adult offspring

The Immune System in Neuropsychiatric Disorders

Immune Activation in the Human Brain

Contrary to the popular myth of the brain as an "immune privileged organ," the nervous and immune systems are significantly interconnected. In fact, cytokines that are traditionally thought of as immune signaling molecules play important roles in neuronal functions as well (for review see [59]). For example, IL-6 modulates neuronal long-term potentiation, a process thought to underlie learning and memory [57, 67]. Cytokines can alter the morphology of several cell types in the brain; neurons retract their processes in response to application of different cytokines, and astroglia and microglia activate and take on different morphology and function after immune stimulation [58, 59]. At the level of the organism, cytokines have been shown to mediate several behaviors. IL-6 and IL-1β are responsible for sickness behavior in mice, defined as the characteristic ruffling of fur, anhedonia, and suppression of appetite [68]. These same cytokines probably cause the feeling of malaise associated with sickness in humans as well. Injection of LPS into human volunteers causes depression- and anxiety-like symptoms [69]. Interferon-alpha (IFN-α) given to cancer patients to lessen the side effects of chemotherapy also causes depression [70]. Clearly, cytokines have the potential to alter behavior in both the developing and adult brain (for review see [71]).

It is possible that abnormal inflammation during gestation and after birth is a core feature of schizophrenia and autism, as there is evidence of ongoing immune activation in the brains of affected individuals. Immune dysregulation has been frequently reported in blood [72–76], cerebrospinal fluid [76, 77], and brain [78] from both schizophrenic (for review see [79]) and autistic individuals (Table 7.2). Moreover, microarray and proteome studies consistently show dysregulation of immune-related transcripts and proteins in schizophrenic [80–82] and autistic [83] brains. The most visually compelling data supporting immune dysregulation is the large amount of glial activation reported in the postmortem autistic brain [78]. Despite different ages, causes of death and comorbid diseases (such as epilepsy) among subjects, a consistent histological pattern of dense, activated microglia (mainly in gray matter) and astroglia (mainly in white matter) was observed. A replication of the microgliosis seen in adult specimens was reported in autistic brains as young as 4-years old [84].

Inflammation and Drug Treatment

Given the biological activity of cytokines, it is likely that immune dysregulation plays an acute as well as developmental role in the behavioral abnormalities of patients. Parents report transient improvement in symptoms when an autistic child is sick [85], and the psychotic symptoms of schizophrenics are also reported to improve

Table 7.2 Inflammatory biomarkers are increased in autism

Sample analyzed	Findings	References
Serum from 23 autistic children	Increased IL-2 and T8 antigen in a subset of children	[74]
Serum from 35 autistics	Increased TNFR II in a subset of children	[76]
Unstimulated whole blood culture supernatant from autistic children	Increased IFNγ and IL-1RA IL-6 trended higher ($p=0.06$)	[75]
Unstimulated PBMCs culture supernatant from 10 autistic children	Increased TH2 cytokines, increased IL-13/IL-10 and IFNγ/IL-10 ratios	[115]
Whole blood from 31 autistic children	Increased monocyte count, increased neopterin level	[116]
CSF from 12 autistic children	Increased biopterin, decreased quinolinic acid and neopterin	[76]
CSF from 10 autistic children	Increased TNFα	[77]
CSF from 6 autistic children	Increases in many cytokines, including IFNγ, TGFβ2 MCP-1 IL-8	[78]
Fresh-frozen tissue from 7 autistic brains	Increases in many cytokines, including MCP-1, IL-6, and TGFβ1	[78]
LPS-stimulated PBMCs from 71 autistic children	Increased TNFα, IL-1β, and IL-6 production	[117]

Studies of serum, blood cell cultures, cerebrospinal fluid (CSF), and brain tissue consistently find elevated levels of proinflammatory markers in autistic individuals. For a thorough review of similar findings in schizophrenia, see [78]

PBMC peripheral blood mononuclear cells; *TNF* tumor necrosis factor; *IFN* interferon; *IL* interleukin; *MCP* monocyte chemoattractant protein; *TNFR* TNF receptor; *IL-1RA* IL-1 receptor antagonist

transiently when those patients have been given experimental malaria in the past [86, 87]. These types of sicknesses would alter the cytokine balance peripherally, and could therefore affect the brain's immune status as well [88]. In addition, there is growing recognition that antipsychotic drugs can have anti-inflammatory effects (for review see [79]). For example, olanzapine, risperidone, perospirone, ziprasidone, quetiapine, and aripiprazole suppress LPS-induced nitric oxide (NO) production in microglial cells [89–91], and some of these drugs suppress proinflammatory cytokines in LPS-treated mice [92]. Perhaps the antipsychotic effects of these drugs are due to their anti-inflammatory properties as well as their antagonism of dopamine type 2 receptors. In fact, several anti-inflammatory drugs have shown promise in the treatment of neuropsychiatric disease. Celecoxib, a COX-2 inhibitor, and minocycline, a tetracycline derivative with anti-inflammatory effects on microglia, have been reported to improve symptoms in schizophrenic patients [93, 94]. The minocycline trial was open-label, while the Celecoxib trial was double-blind and placebo-controlled. Pioglitazone (Actos©), also an anti-inflammatory drug that inhibits microglial activation, showed promise in an early trial in autistic patients [95]. Parents reported improvements in several core behaviors in 76% of 25 children enrolled. An open label trial of minocycline for the treatment of autism is ongoing at the NIH at the time of this writing. However, trials testing drugs in children are time-consuming, expensive, and ethically delicate. Novel treatments based on anti-inflammatory therapies can initially be tested in the rodent models described here that have experimental validity.

Mice born to dams that received immune-activating treatments during pregnancy also show increased levels of inflammation in both peripheral organs and the brain. For example, three injections of IL-6 into pregnant mice produced elevated levels of IL-6 in the serum of the adult offspring [66]. In addition, two other animal models meant to mimic early insults relevant to autism, injection of antibodies from mothers of autistic children into pregnant mice and injection of terbutaline into neonatal rats, both produce microglial activation in the brains of adult animals [96, 97]. These two examples demonstrate the ability of various animal models to replicate immune activation in neuropsychiatric disease. Daily maternal injection of LPS throughout pregnancy produces offspring with chronically elevated levels of peripheral inflammatory cytokines. Remarkably, treatment with antipsychotic drugs normalizes not only their behavior, but also their cytokine profiles [98]. These animal models highlight the utility of rodent models, particularly MIA, for features of human neuropsychiatric disease and demonstrate the power of antipsychotic drugs to dampen abnormal inflammation while altering behavior.

Mechanisms for Chronic Inflammation and Abnormal Behavior

How does immune activation early in development set up an ongoing inflammatory state that persists months or years after the original insult? Does immune activation cause neuropsychiatric disorders, or does a mis-wired brain cause inflammation? Ongoing research is just beginning to address these questions. In the meantime, analogies to other, more thoroughly characterized model systems may guide our thinking.

One hypothesis proposes that the abnormal immune system precipitates abnormal behavior in the offspring, similar to a model of maternal stress. Maternal stress causes persistent behavioral, physiological, and biochemical changes in the adult offspring. Maternal stress is a risk factor for schizophrenia [55], and a rat model was developed in the mid-1990s in which researchers stressed pregnant rats by restraining them in a tube for several hours each day. The offspring display an elevated response to stress as adults (as well as other abnormal behaviors). When stressed, the offspring of stressed mothers show a prolonged elevation of corticosterone after the stressor is removed compared with offspring of unstressed rats [56]. A similar result was recently reported in people whose mothers were exposed to stress during pregnancy [99]. This effect is dependent on corticosteroids, as stress in adrenalectomized pregnant rats does not affect the offspring, while corticosteroid injection does [56]. Mechanistically, corticosteroid injections alter the number of glucocorticoid receptors in the hippocampus, an area important in terminating the stress response [100, 101]. A related model involves the stress response of young rats whose mothers show high or low levels of pup-directed licking and grooming behavior (LG). High-LG offspring have a reduced stress response compared with low-LG offspring, with correspondingly different levels of glucocorticoid receptor expression. Maternal behavior alters glucocorticoid receptor expression in the offspring by altering histone acetyltransferase activity and subsequent methylation of the glucocorticoid receptor gene (for review see [102]). By analogy, perhaps a similar cytokine "homeostat" is set early in development based on the low-level cytokine exposure that is essential for normal brain development. In the MIA model where cytokines are abnormally elevated during a critical period of development, levels of a receptor protein could be altered, perhaps via deoxy-ribonucleic acid (DNA) methylation, leading to persistent changes in baseline cytokine levels.

An alternative hypothesis suggests that the abnormal brain causes an abnormal immune system. If, for example, inhibitory gamma-amino butyric acid (GABA-ergic) interneurons fail to properly develop, the resulting brain circuitry would show an imbalance between levels of excitation and inhibition, favoring excitation. Clinically, this may manifest as seizures, as observed in 30–40% of autistic children, or the loss of GABAergic markers such as (PV), as is seen in schizophrenic brains [103, 104], and rodent offspring of MIA mothers [33]. Psychomimetic drugs such as ketamine, an *N*-methyl-D-aspartate (NMDA) antagonist, produce schizophrenia-like symptoms by inhibiting GABAergic neurons and producing a dis-inhibition of excitatory circuits [105]. In fact, loss of GABAergic signaling may be a core feature of schizophrenia [106]. Hyperexcited neuronal circuits can also produce an inflammatory state. Using a mouse model in which subchronic administration of ketamine induced a loss of PV-positive neurons, Behrens et al. [107] demonstrated that ketamine caused long-term damage to PV-positive neurons by activating the superoxide-producing enzyme NADPH oxidase (NOX). NOX production was, in turn, dependent on neuronal IL-6 produced in response to sustained hyperexcitability [105]. This activation could be a homeostatic response to reduced inhibition whereby IL-6-dependent free radical production reduces excitatory transmission (via sensitivity of glutamatergic synapses to free radicals) and restores the inhibitory/excitatory balance of the circuit [105]. However, prolonged exposure to ketamine overactivates this feedback system, and the free radical production causes long-term damage to PV-positive interneurons. Similarly, if IL-6 elevation due to MIA were to reach the fetus at a critical period of development, it could cause a reduction in PV interneurons, as reported in the MIA model [33]. This loss of inhibition could create a hyperexcited circuit similar to a ketamine-exposed brain, leading to more IL-6 production by a mechanism similar to ketamine exposure. This endogenous IL-6 production would further suppress PV-positive interneurons, creating more excitation and an ongoing positive feedback loop. Perhaps this process is set into motion prenatally and continues after birth, resulting in autism, or perhaps a marginally stable circuit needs some later stimulus, such as peri-pubertal synaptic pruning, to tip the balance and create an inflammatory, hyperexcited, hallucinatory state such as in schizophrenia.

Conclusion

Prenatal infections are now a well-established risk factor for later development of neuropsychiatric disease. Animal models of these risk factors based on maternal immune activation have demonstrated that early life exposure to cytokines, particularly IL-6, can have long-term affects on both the immune status and the normal functioning of the brain. Recognizing this interrelatedness of the brain and the immune system allows us to look at neuropsychiatric illness differently. Emerging evidence including the behavioral consequences of early life exposure to inflammatory stimuli, the profound dysregulation of the brain's immune status seen in neuropsychiatric illness, and the anti-inflammatory action of antipsychotic drugs (and the potential efficacy of anti-inflammatory drugs as antipsychotics) suggests that a dysregulated immune system is central to neuropsychiatric disease. Whether an affected brain causes inflammation, or inflammation causes a brain to malfunction, is an important issue in current research. Ongoing animal and human drug trails investigating treatment of neuropsychiatric diseases with anti-inflammatory drugs will also yield critical clues to the nature of this relationship. The neuro-immune model of neuropsychiatric disease could fundamentally alter how we diagnose, treat, and prevent these disorders.

Acknowledgments We thank Wensi Sheng for designing Fig. 7.1. Work from the authors' laboratory received support from the National Association for Autism Research, the NIMH, and the McKnight Foundation.

References

1. Palomo R, Belinchon M, Ozonoff S (2006) Autism and family home movies: a comprehensive review. J Dev Behav Pediatr 27(2 Suppl):S59–S68.
2. Lewis DA, Levitt P (2002) Schizophrenia as a disorder of neurodevelopment. Annu Rev Neurosci 25:409–432.
3. Lewis DA, Lieberman JA (2000) Catching up on schizophrenia: natural history and neurobiology. Neuron 28(2):325–334.
4. Sykes NH, Lamb JA (2007) Autism: the quest for the genes. Expert Rev Mol Med 9(24):1–15.
5. Grice DE, Buxbaum JD (2006) The genetics of autism spectrum disorders. Neuromolecular Med 8(4):451–460.
6. Cook EH Jr, Scherer SW (2008) Copy-number variations associated with neuropsychiatric conditions. Nature 455(7215):919–923.
7. Schanen NC (2006) Epigenetics of autism spectrum disorders. Hum Mol Genet 15(2):138–150.
8. Trottier G, Srivastava L, Walker CD (1999) Etiology of infantile autism: a review of recent advances in genetic and neurobiological research. J Psychiatry Neurosci 24(2):103–115.
9. Alsdorf R, Wyszynski DF (2005) Teratogenicity of sodium valproate. Expert Opin Drug Saf 4(2):345–353.
10. Brown AS (2006) Prenatal infection as a risk factor for schizophrenia. Schizophr Bull 32(2):200–202.

11. Patterson PH (2002) Maternal infection: window on neuroimmune interactions in fetal brain development and mental illness. Curr Opin Neurobiol 12(1):115–118.
12. Chess S (1977) Follow-up report on autism in congenital rubella. J Autism Child Schizophr 7(1):69–81.
13. Mednick SA, Machon RA, Huttunen MO, Bonett D (1988) Adult schizophrenia following prenatal exposure to an influenza epidemic. Arch Gen Psychiatry 45(2):189–192.
14. Bagalkote H, Pang D, Jones P (2001) Maternal influenza and schizophrenia in the offspring. Int J Ment Health 39:3–21.
15. Brown AS, Begg MD, Gravenstein S et al (2004) Serologic evidence of prenatal influenza in the etiology of schizophrenia. Arch Gen Psychiatry 61(8):774–780.
16. Byrne M, Agerbo E, Bennedsen B, Eaton WW, Mortensen PB (2007) Obstetric conditions and risk of first admission with schizophrenia: a Danish national register based study. Schizophr Res 97(1–3):51–59.
17. Brown AS, Hooton J, Schaefer CA et al (2004) Elevated maternal interleukin-8 levels and risk of schizophrenia in adult offspring. Am J Psychiatry 161(5):889–895.
18. Brown AS, Schaefer CA, Quesenberry CP Jr, Liu L, Babulas VP, Susser ES (2005) Maternal exposure to toxoplasmosis and risk of schizophrenia in adult offspring. Am J Psychiatry 162(4):767–773.
19. Mortensen PB, Norgaard-Pedersen B, Waltoft BL, Sorensen TL, Hougaard D, Yolken RH (2007) Early infections of Toxoplasma gondii and the later development of schizophrenia. Schizophr Bull 33(3):741–744.
20. Buka SL, Tsuang MT, Torrey EF, Klebanoff MA, Bernstein D, Yolken RH (2001) Maternal infections and subsequent psychosis among offspring. Arch Gen Psychiatry 58(11):1032–1037.
21. Babulas V, Factor-Litvak P, Goetz R, Schaefer CA, Brown AS (2006) Prenatal exposure to maternal genital and reproductive infections and adult schizophrenia. Am J Psychiatry 163(5):927–929.
22. Sorensen HJ, Mortensen EL, Reinisch JM, Mednick SA (2009) Association between prenatal exposure to bacterial infection and risk of schizophrenia. Schizophr Bull 35(3):631–637.
23. Brown AS, Deicken RF, Vinogradov S et al (2009) Prenatal infection and cavum septum pellucidum in adult schizophrenia. Schizophr Res 108:104–113.
24. Brown AS. (2009) Prenatal infection and executive dysfunction in adult schizophrenia. Am J Psychiatry 166(6):631–634.
25. Desmond MM, Wilson GS, Melnick JL et al (1967) Congenital rubella encephalitis. Course and early sequelae. J Pediatr 71(3):311–331.
26. Ciaranello AL, Ciaranello RD (1995) The neurobiology of infantile autism. Annu Rev Neurosci 18:101–128.
27. Brown AS, Susser ES (2002) In utero infection and adult schizophrenia. Ment Retard Dev Disabil Res Rev 8(1):51–57.
28. Shi L, Fatemi SH, Sidwell RW, Patterson PH (2003) Maternal influenza infection causes marked behavioral and pharmacological changes in the offspring. J Neurosci 23(1):297–302.
29. Wynn JK, Dawson ME, Schell AM, McGee M, Salveson D, Green MF (2004) Prepulse facilitation and prepulse inhibition in schizophrenia patients and their unaffected siblings. Biol Psychiatry 55(5):518–523.
30. Turetsky BI, Calkins ME, Light GA, Olincy A, Radant AD, Swerdlow NR (2007) Neurophysiological endophenotypes of schizophrenia: the viability of selected candidate measures. Schizophr Bull 33(1):69–94.
31. Perry W, Minassian A, Lopez B, Maron L, Lincoln A (2007) Sensorimotor gating deficits in adults with autism. Biol Psychiatry 61(4):482–486.
32. Fatemi SH, Emamian ES, Kist D et al (1999) Defective corticogenesis and reduction in Reelin immunoreactivity in cortex and hippocampus of prenatally infected neonatal mice. Mol Psychiatry 4(2):145–154.

33. Meyer U, Nyffeler M, Yee BK, Knuesel I, Feldon J (2008) Adult brain and behavioral pathological markers of prenatal immune challenge during early/middle and late fetal development in mice. Brain Behav Immun 22(4):469–486.
34. Shi L, Smith SE, Malkova N, Tse D, Patterson PH (2008) Activation of the maternal immune system alters cerebellar development in the offspring. Brain Behav Immun 23(1):116–123.
35. Amaral DG, Schumann CM, Nordahl CW (2008) Neuroanatomy of autism. Trends Neurosci 31(3):137–145.
36. Taubenberger JK, Morens DM (2008) The pathology of influenza virus infections. Annu Rev Pathol 3:499–522.
37. Shi L, Tu N, Patterson PH (2005) Maternal influenza infection is likely to alter fetal brain development indirectly: the virus is not detected in the fetus. Int J Dev Neurosci 23(2–3):299–305.
38. Smith SE, Li J, Garbett K, Mirnics K, Patterson PH (2007) Maternal immune activation alters fetal brain development through interleukin-6. J Neurosci 27(40):10695–10702.
39. Piontkewitz Y, Weiner I, Assaf Y (2007) Post-pubertal emergence of schizophrenia-like abnormalaties following prenatal immune system activation and thier prevention: modeling the disorder and its prodrome. In: 7th IBRO World Congress of Neuroscience, Melbourne, Australia, 45.
40. Zuckerman L, Rehavi M, Nachman R, Weiner I (2003) Immune activation during pregnancy in rats leads to a postpubertal emergence of disrupted latent inhibition, dopaminergic hyperfunction, and altered limbic morphology in the offspring: a novel neurodevelopmental model of schizophrenia. Neuropsychopharmacology 28(10):1778–1789.
41. Ozawa K, Hashimoto K, Kishimoto T, Shimizu E, Ishikura H, Iyo M (2006) Immune activation during pregnancy in mice leads to dopaminergic hyperfunction and cognitive impairment in the offspring: a neurodevelopmental animal model of schizophrenia. Biol Psychiatry 59(6):546–554.
42. Nyffeler M, Meyer U, Yee BK, Feldon J, Knuesel I (2006) Maternal immune activation during pregnancy increases limbic $GABA_A$ receptor immunoreactivity in the adult offspring: implications for schizophrenia. Neuroscience 143(1):51–62.
43. Romero R, Gotsch F, Pineles B, Kusanovic JP (2007) Inflammation in pregnancy: its roles in reproductive physiology, obstetrical complications, and fetal injury. Nutr Rev 65(12 Pt 2): S194–S202.
44. O'Callaghan E, Sham PC, Takei N et al (1994) The relationship of schizophrenic births to 16 infectious diseases. Br J Psychiatry 165(3):353–356.
45. Watson CG, Kucala T, Tilleskjor C, Jacobs L (1984) Schizophrenic birth seasonality in relation to the incidence of infectious diseases and temperature extremes. Arch Gen Psychiatry 41(1):85–90.
46. Borrell J, Vela JM, Arevalo-Martin A, Molina-Holgado E, Guaza C (2002) Prenatal immune challenge disrupts sensorimotor gating in adult rats. Implications for the etiopathogenesis of schizophrenia. Neuropsychopharmacology 26(2):204–215.
47. Fortier ME, Kent S, Ashdown H, Poole S, Boksa P, Luheshi GN (2004) The viral mimic, polyinosinic:polycytidylic acid, induces fever in rats via an interleukin-1-dependent mechanism. Am J Physiol 287(4):R759–R766.
48. Golan HM, Lev V, Hallak M, Sorokin Y, Huleihel M (2005) Specific neurodevelopmental damage in mice offspring following maternal inflammation during pregnancy. Neuropharmacology 48(6):903–917.
49. Hava G, Vered L, Yael M, Mordechai H, Mahoud H (2006) Alterations in behavior in adult offspring mice following maternal inflammation during pregnancy. Dev Psychobiol 48(2):162–168.
50. Cai Z, Pan ZL, Pang Y, Evans OB, Rhodes PG (2000) Cytokine induction in fetal rat brains and brain injury in neonatal rats after maternal lipopolysaccharide administration. Pediatr Res 47(1):64–72.
51. Ling Z, Chang QA, Tong CW, Leurgans SE, Lipton JW, Carvey PM (2004) Rotenone potentiates dopamine neuron loss in animals exposed to lipopolysaccharide prenatally. Exp Neurol 190(2):373–383.

52. Paintlia MK, Paintlia AS, Barbosa E, Singh I, Singh AK (2004) N-acetylcysteine prevents endotoxin-induced degeneration of oligodendrocyte progenitors and hypomyelination in developing rat brain. J Neurosci Res 78(3):347–361.
53. Bobetsis YA, Barros SP, Offenbacher S (2006) Exploring the relationship between periodontal disease and pregnancy complications. J Am Dent Assoc 137:7S–13S.
54. Wang X, Rousset CI, Hagberg H, Mallard C (2006) Lipopolysaccharide-induced inflammation and perinatal brain injury. Semin Fetal Neonatal Med 11(5):343–353.
55. Beydoun H, Saftlas AF (2008) Physical and mental health outcomes of prenatal maternal stress in human and animal studies: a review of recent evidence. Paediatr Perinat Epidemiol 22(5):438–466.
56. Barbazanges A, Piazza PV, Le Moal M, Maccari S (1996) Maternal glucocorticoid secretion mediates long-term effects of prenatal stress. J Neurosci 16(12):3943–3949.
57. Jankowsky JL, Patterson PH (1999) Cytokine and growth factor involvement in long-term potentiation. Mol Cell Neurosci 14(6):273–286.
58. Gilmore JH, Jarskog LF, Vadlamudi S, Lauder J (2004) Prenatal infection and risk for schizophrenia: IL-I beta, IL-6, and TNF alpha inhibit cortical neuron dendrite development. Neuropsychopharmacology 29(7):1221–1229.
59. Bauer S, Kerr BJ, Patterson PH (2007) The neuropoietic cytokine family in development, plasticity, disease and injury. Nat Rev 8(3):221–232.
60. Dahlgren J, Samuelsson AM, Jansson T, Holmang A (2006) Interleukin-6 in the maternal circulation reaches the rat fetus in mid-gestation. Pediatr Res 60(2):147–151.
61. Ponzio NM, Servatius R, Beck K, Marzouk A, Kreider T (2007) Cytokine levels during pregnancy influence immunological profiles and neurobehavioral patterns of the offspring. Ann N Y Acad Sci 1107:118–128.
62. Meyer U, Murray PJ, Urwyler A, Yee BK, Schedlowski M, Feldon J (2007) Adult behavioral and pharmacological dysfunctions following disruption of the fetal brain balance between pro-inflammatory and IL-10-mediated anti-inflammatory signaling. Mol Psychiatry 13:208–221.
63. Pang Y, Rodts-Palenik S, Cai Z, Bennett WA, Rhodes PG (2005) Suppression of glial activation is involved in the protection of IL-10 on maternal E. coli induced neonatal white matter injury. Brain Res Dev Brain Res 157(2):141–149.
64. Silver RM, Lohner WS, Daynes RA, Mitchell MD, Branch DW (1994) Lipopolysaccharide-induced fetal death: the role of tumor-necrosis factor alpha. Biol Reprod 50(5):1108–1112.
65. Xu DX, Chen YH, Wang H, Zhao L, Wang JP, Wei W (2006) Tumor necrosis factor alpha partially contributes to lipopolysaccharide-induced intra-uterine fetal growth restriction and skeletal development retardation in mice. Toxicol Lett 163(1):20–29.
66. Samuelsson AM, Jennische E, Hansson HA, Holmang A (2006) Prenatal exposure to interleukin-6 results in inflammatory neurodegeneration in hippocampus with NMDA/GABA($_A$) dysregulation and impaired spatial learning. Am J Physiol 290(5):R1345–R1356.
67. Balschun D, Wetzel W, Del Rey A et al (2004) Interleukin-6: a cytokine to forget. FASEB J 18(14):1788–1790.
68. Swiergiel AH, Smagin GN, Johnson LJ, Dunn AJ (1997) The role of cytokines in the behavioral responses to endotoxin and influenza virus infection in mice: effects of acute and chronic administration of the interleukin-1-receptor antagonist (IL-1ra). Brain Res 776(1–2):96–104.
69. Reichenberg A, Yirmiya R, Schuld A et al (2001) Cytokine-associated emotional and cognitive disturbances in humans. Arch Gen Psychiatry 58(5):445–452.
70. Raison CL, Capuron L, Miller AH (2006) Cytokines sing the blues: inflammation and the pathogenesis of depression. Trends Immunol 27(1):24–31.
71. Yirmiya R (2000) Depression in medical illness: the role of the immune system. West J Med 173(5):333–336.
72. Garver DL, Tamas RL, Holcomb JA (2003) Elevated interleukin-6 in the cerebrospinal fluid of a previously delineated schizophrenia subtype. Neuropsychopharmacology 28(8): 1515–1520.

73. Zhang XY, Zhou DF, Cao LY, Wu GY, Shen YC (2005) Cortisol and cytokines in chronic and treatment-resistant patients with schizophrenia: association with psychopathology and response to antipsychotics. Neuropsychopharmacology 30(8):1532–1538.
74. Singh VK, Warren RP, Odell JD, Cole P (1991) Changes of soluble interleukin-2, interleukin-2 receptor, T8 antigen, and interleukin-1 in the serum of autistic children. Clin Immunol Immunopathol 61(3):448–455.
75. Croonenberghs J, Bosmans E, Deboutte D, Kenis G, Maes M (2002) Activation of the inflammatory response system in autism. Neuropsychobiology 45(1):1–6.
76. Zimmerman AW, Jyonouchi H, Comi AM et al (2005) Cerebrospinal fluid and serum markers of inflammation in autism. Pediatr Neurol 33(3):195–201.
77. Chez MG, Dowling T, Patel PB, Khanna P, Kominsky M (2007) Elevation of tumor necrosis factor-alpha in cerebrospinal fluid of autistic children. Pediatr Neurol 36(6):361–365.
78. Vargas DL, Nascimbene C, Krishnan C, Zimmerman AW, Pardo CA (2005) Neuroglial activation and neuroinflammation in the brain of patients with autism. Ann Neurol 57(1):67–81.
79. Drzyzga L, Obuchowicz E, Marcinowska A, Herman ZS (2006) Cytokines in schizophrenia and the effects of antipsychotic drugs. Brain Behav Immun 20(6):532–545.
80. Martins-de-Souza D, Gattaz WF, Schmitt A et al (2009) Prefrontal cortex shotgun proteome analysis reveals altered calcium homeostasis and immune system imbalance in schizophrenia. Eur Arch Psychiatry Clin Neurosci 259:151–163.
81. Arion D, Unger T, Lewis DA, Levitt P, Mirnics K (2007) Molecular evidence for increased expression of genes related to immune and chaperone function in the prefrontal cortex in schizophrenia. Biol Psychiatry 62(7):711–721.
82. Saetre P, Emilsson L, Axelsson E, Kreuger J, Lindholm E, Jazin E (2007) Inflammation-related genes up-regulated in schizophrenia brains. BMC Psychiatry 7:46.
83. Garbett K, Ebert PJ, Mitchell A et al (2008) Immune transcriptome alterations in the temporal cortex of subjects with autism. Neurobiol Dis 30(3):303–311.
84. Morgan JT, Chana G, Buckwalter J, Courchesne E, Everall IP (2007) Increased Iba-1 positive microglial cell density in the autistic brain. Society for Neuroscience poster presentation.
85. Curran LK, Newschaffer CJ, Lee LC, Crawford SO, Johnston MV, Zimmerman AW (2007) Behaviors associated with fever in children with autism spectrum disorders. Pediatrics 120(6):e1386–e1392.
86. Hinsie LE (1929) Malaria treatment of schizophrenia. Psychiatr Q 3:210–214.
87. Tempelton WL, Glas CB (1924) The effect of malarial fever upon dementia praecox subjects. J Ment Sci 70:92–95.
88. Dantzer R, Kelley KW (2007) Twenty years of research on cytokine-induced sickness behavior. Brain Behav Immun 21(2):153–160.
89. Hou Y, Wu CF, Yang JY et al (2006) Effects of clozapine, olanzapine and haloperidol on nitric oxide production by lipopolysaccharide-activated N9 cells. Prog Neuropsychopharmacol Biol Psychiatry 30(8):1523–1528.
90. Kato T, Monji A, Hashioka S, Kanba S (2007) Risperidone significantly inhibits interferon-gamma-induced microglial activation in vitro. Schizophr Res 92(1–3):108–115.
91. Bian Q, Kato T, Monji A et al (2008) The effect of atypical antipsychotics, perospirone, ziprasidone and quetiapine on microglial activation induced by interferon-gamma. Prog Neuropsychopharmacol Biol Psychiatry 32(1):42–48.
92. Sugino H, Futamura T, Mitsumoto Y, Maeda K, Marunaka Y (2009) Atypical antipsychotics suppress production of proinflammatory cytokines and up-regulate interleukin-10 in lipopolysaccharide-treated mice. Prog Neuropsychopharmacol Biol Psychiatry 33(2): 303–307.
93. Akhondzadeh S, Tabatabaee M, Amini H, Ahmadi Abhari SA, Abbasi SH, Behnam B (2007) Celecoxib as adjunctive therapy in schizophrenia: a double-blind, randomized and placebo-controlled trial. Schizophr Res 90(1–3):179–185.

94. Miyaoka T, Yasukawa R, Yasuda H, Hayashida M, Inagaki T, Horiguchi J (2008) Minocycline as adjunctive therapy for schizophrenia: an open-label study. Clin Neuropharmacol 31(5):287–292.
95. Boris M, Kaiser CC, Goldblatt A et al (2007) Effect of pioglitazone treatment on behavioral symptoms in autistic children. J Neuroinflammation 4:3.
96. Zerrate MC, Pletnikov M, Connors SL et al (2007) Neuroinflammation and behavioral abnormalities after neonatal terbutaline treatment in rats: implications for autism. J Pharmacol Exp Ther 322(1):16–22.
97. Singer HS, Morris C, Gause C, Pollard M, Zimmerman AW, Pletnikov M (2009) Prenatal exposure to antibodies from mothers of children with autism produces neurobehavioral alterations: a pregnant dam mouse model. J Neuroimmunol 211(1–2):39–48.
98. Romero E, Ali C, Molina-Holgado E, Castellano B, Guaza C, Borrell J (2007) Neurobehavioral and immunological consequences of prenatal immune activation in rats. Influence of antipsychotics. Neuropsychopharmacology 32(8):1791–1804.
99. Entringer S, Kumsta R, Hellhammer DH, Wadhwa PD, Wust S (2009) Prenatal exposure to maternal psychosocial stress and HPA axis regulation in young adults. Horm Behav 55(2):292–298.
100. Meaney MJ, Diorio J, Francis D et al (1996) Early environmental regulation of forebrain glucocorticoid receptor gene expression: implications for adrenocortical responses to stress. Dev Neurosci 18(1–2):49–72.
101. Levitt NS, Lindsay RS, Holmes MC, Seckl JR (1996) Dexamethasone in the last week of pregnancy attenuates hippocampal glucocorticoid receptor gene expression and elevates blood pressure in the adult offspring in the rat. Neuroendocrinology 64(6):412–418.
102. Weaver IC (2009) Epigenetic effects of glucocorticoids. Semin Fetal Neonatal Med 14(3):143–150.
103. Beasley CL, Reynolds GP (1997) Parvalbumin-immunoreactive neurons are reduced in the prefrontal cortex of schizophrenics. Schizophr Res 24(3):349–355.
104. Hashimoto T, Volk DW, Eggan SM et al (2003) Gene expression deficits in a subclass of GABA neurons in the prefrontal cortex of subjects with schizophrenia. J Neurosci 23(15):6315–6326.
105. Behrens MM, Ali SS, Dugan LL (2008) Interleukin-6 mediates the increase in NADPH-oxidase in the ketamine model of schizophrenia. J Neurosci 28(51):13957–13966.
106. Lewis DA, Hashimoto T, Volk DW (2005) Cortical inhibitory neurons and schizophrenia. Nat Rev 6(4):312–324.
107. Behrens MM, Ali SS, Dao DN et al (2007) Ketamine-induced loss of phenotype of fast-spiking interneurons is mediated by NADPH-oxidase. Science 318(5856): 1645–1647.
108. Ashdown H, Dumont Y, Ng M, Poole S, Boksa P, Luheshi GN (2006) The role of cytokines in mediating effects of prenatal infection on the fetus: implications for schizophrenia. Mol Psychiatry 11(1):47–55.
109. Golan H, Stilman M, Lev V, Huleihel M (2006) Normal aging of offspring mice of mothers with induced inflammation during pregnancy. Neuroscience 141(4):1909–1918.
110. Liverman CS, Kaftan HA, Cui L et al (2006) Altered expression of pro-inflammatory and developmental genes in the fetal brain in a mouse model of maternal infection. Neurosci Lett 399(3):220–225.
111. Urakubo A, Jarskog LF, Lieberman JA, Gilmore JH (2001) Prenatal exposure to maternal infection alters cytokine expression in the placenta, amniotic fluid, and fetal brain. Schizophr Res 47(1):27–36.
112. Gilmore JH, Jarskog LF, Vadlamudi S (2005) Maternal poly I:C exposure during pregnancy regulates TNF alpha, BDNF, and NGF expression in neonatal brain and the maternal-fetal unit of the rat. J Neuroimmunol 159(1–2):106–112.

113. Meyer U, Nyffeler M, Engler A et al (2006) The time of prenatal immune challenge determines the specificity of inflammation-mediated brain and behavioral pathology. J Neurosci 26(18):4752–4762.
114. Meyer U, Feldon J, Schedlowski M, Yee BK (2006) Immunological stress at the maternal-foetal interface: a link between neurodevelopment and adult psychopathology. Brain Behav Immun 20(4):378–388.
115. Molloy CA, Morrow AL, Meinzen-Derr J et al (2006) Elevated cytokine levels in children with autism spectrum disorder. J Neuroimmunol 172(1–2):198–205.
116. Sweeten TL, Posey DJ, McDougle CJ (2003) High blood monocyte counts and neopterin levels in children with autistic disorder. Am J Psychiatry 160(9):1691–1693.
117. Jyonouchi H, Sun S, Le H (2001) Proinflammatory and regulatory cytokine production associated with innate and adaptive immune responses in children with autism spectrum disorders and developmental regression. J Neuroimmunol 120(1–2):170–179.

Chapter 8
Prenatal Infections and Schizophrenia in Later Life – Focus on *Toxoplasma gondii*

Robert Yolken and E. Fuller Torrey

Keywords Prenatal maternal infection • *Toxoplasma gondii* • *Toxoplasma* antibodies • Serotypes • Herpesviruses • Schizophrenia

Introduction

Schizophrenia and bipolar disorder are prevalent neuropsychiatric disorders that are major causes of morbidity and mortality in the United States and most other areas of the world. The etiology of these disorders remains obscure. Family studies have indicated that both have a high degree of heritability [1, 2]. This has led to an extensive search for genes that cause these diseases. However, despite extensive studies, genetic methods have failed to identify genes that have a major effect on disease risk in a large number of individuals [3]. While copy number variants that have a strong association with schizophrenia in some cases have been identified [4], these variants appear to be rare in the general population and cannot explain disease risk in a large number of individuals [5]. On the other hand, common variants have been identified that have a statistical association with disease risk. However, these have generally been associated with relatively low odds ratios (<1.5) and are thus not highly predictive of disease risk [6].

Because of these limitations, a number of investigators have suggested alternate mechanisms to explain the high degree of familial association with risk of schizophrenia. One possible mechanism is that shared risk among family members may be related to a common exposure to an infectious agent or other environmental

R. Yolken (✉)
Stanley Division of Developmental Neurovirology, Johns Hopkins School of Medicine, 600 N. Wolfe Street, Blalock 1105, Baltimore, MD, 21287-4933, USA
e-mail: yolken@mail.jhmi.edu

A.W. Zimmerman and S.L. Connors (eds.), *Maternal Influences on Fetal Neurodevelopment: Clinical and Research Aspects*,
DOI 10.1007/978-1-60327-921-5_8, © Springer Science+Business Media, LLC 2010

factor during fetal development or during the neonatal period. The plausibility of this scenario is supported by numerous epidemiological studies indicating that exposures during pregnancy and early life confer an increased risk of schizophrenia in later life. Exposures in utero that have been associated with an increased risk of this disorder include nutritional deficiencies, preeclampsia and other perinatal complications, fever, medications, and exposure to infectious agents [7–9]. In addition, studies have documented that an increased risk of schizophrenia is associated with demographic factors which are likely to be associated with infectious disease exposure during pregnancy and early life. These include season of birth [10], household crowding [11], urban birth [12], and immigration [13].

Recent advances in methods for the detection and characterization of infectious diseases have led to renewed interest in studying the role of specific infectious agents and psychiatric disorders such as schizophrenia. These methods include ones that allow for the large-scale measurement of antibodies to infectious agents in small volumes of blood samples obtained from individuals in cohort studies and stored for many years. Furthermore, there is an increased understanding of the interaction of infectious agents and host genetic factors, thus allowing for the incorporation of exposure to infectious agents into genetic models of disease risk [14].

Specific infectious agents and disease processes during fetal development and early life that have been associated with an increased risk of schizophrenia are shown in Table 8.1. This chapter will examine some of the methodological issues involved in the measurement of these associations and examine in detail the role of the protozoan parasite *Toxoplasma gondii* in schizophrenia. This protozoan parasite is selected for detailed analysis in light of multiple epidemiological studies documenting its association with schizophrenia as well as an emerging body of information relating to mechanisms by which this organism can infect the brain and alter behavior.

Methods for the Study of Perinatal Infections and Risk of Schizophrenia

Case Control Studies

One challenge regarding the association between exposures during fetal development and early life and the subsequent development of adult psychiatric disorders is the design of studies to measure this association. A major issue in study design is the long time interval between the measured exposure and the development of symptoms, a period of time which can be longer than 30 years. Initial studies were largely based on case control designs dependent on maternal recall of events occurring during pregnancy and infant development. These studies have been valuable in terms of initially identifying the association between perinatal events as risk factors

Table 8.1 Infectious diseases during the prenatal period and early life associated with schizophrenia in later life

A. Specific infectious agents		
Organism	Positive studies	Negative studies
Toxoplasma gondii	Prospective cohort [37], case registry [27]	Prospective cohort [57]
Herpes simplex virus type 2	Prospective cohort [85], case registry [86]	Prospective cohort [85]
Influenza viruses	Prospective cohort [87, 88], population [89]	Prospective cohort [87] population [25], case registry [90]
Rubella virus	Case control [91]	Prospective cohort [85] population [25]
Polio virus	Population [25], case registry [92]	Population [93]
Measles virus	Population [25]	Population [94]
B. General factors and conditions that might be associated with infectious agents		
Factor	Positive studies	Negative studies
Maternal bacterial infections – respiratory	Case register [95], prospective cohort	
Maternal bacterial infections – genitourinary	Case register [95], prospective cohort [96]	
Viral central nervous system infections (infancy and childhood)	Case register [29, 97]	Case register [98]
Pet cats	Case control [17]	
Environmental lead	Prospective cohort [99]	
Nutritional deficiency	Population [24, 100]	

for the later development of schizophrenia. Case control studies relying on maternal recall have also been useful for the identification of specific risk factors such as perinatal complications [15], exposure to influenza [16], maternal fever, and exposure to pets [17].

Prospective Cohort Studies

Despite the fact that case–control studies have yielded valuable information, these types of studies are subject to a number of biases based on case finding and memory of perinatal events. Even though much maternal recall has been shown to be accurate in many situations [18, 19], it would be preferable to also have studies that are not dependent upon maternal recall. The ideal study in this regard would be a prospective cohort study, where pregnant women were enrolled and their offspring followed until they had completed the age of risk for new-onset schizophrenia or bipolar disorder, which is around 30 years of age. Furthermore, the ideal study would involve the obtaining and storage of blood and other biological samples so that environmental exposures could be measured in an objective fashion. There is currently no study that meets these criteria and from which data and samples related to perinatal exposure and adult schizophrenia are available.

The recently established National Children's Study has this as one of its goals [20]. This study is a 21-year prospective study that will enroll 100,000 children, measure a range of exposures, and follow the children until adulthood. This study should provide extensive information relating to exposures during pregnancy and early life and the risk of a range of psychiatric disorders with onset during childhood and early adulthood. In addition, this study will involve the collection and storage of a wide range of biological samples, which can be analyzed for the presence of infectious agents and other environmental toxins. However, the fact that this study started enrolling participants during the year 2006 means that these data will not be available until around the year 2027. Similar cohort-based studies being performed in Europe hold additional promise for future studies of adult psychiatric disorders [21].

While there have been no previous prospective perinatal cohort studies that have involved the continuous monitoring of associations between early life exposures and adult psychiatric disorders, there were several perinatal cohort studies performed in the past, which have provided important information relating to these associations. The most important of these was the Collaborative Perinatal Study, which started in 1958. In this, more than 55,000 pregnant women were enrolled between 1959 and 1966 at 11 study sites within the United States. The cohort mothers were intensively studied during pregnancy and their infants were evaluated for physical and intellectual development during the first 7 years of life [18, 19]. Data were also obtained on potential environmental exposures such as smoking and exposure to medications. In addition to detailed clinical evaluations, blood samples were obtained from the mothers and from many of the infants in the form of cord blood. These samples were stored in a repository where they are available for future analyses [22]. The demographic and clinical data from the Collaborative Perinatal Study are available from public sources (http://www.archives.gov/research/electronic-records/nih.html) and the blood samples are available to the scientific community for specific studies following peer review procedures. A parallel study was performed at Kaiser Permanente on a smaller cohort [17], although the data and samples from this cohort have generally not as yet been made available to the general scientific community.

While the active collection of data from the participants in these cohorts was stopped in the 1960s in most of the study sites, a number of investigators have been able to recontact the participants and identify individuals with psychiatric disorders. Since blood samples had been obtained from the mothers during pregnancy and from the mothers and the infants at birth, this study has allowed for the direct testing of exposures to environmental factors during pregnancy and the development of schizophrenia and related disorders in later life. Limitations of these cohort studies include the fact that the cases of schizophrenia were found by the recontact of the study individuals or the search of medical records. This may have introduced some biases in terms of the individuals who were available for recontact and reevaluation. Secondly, blood samples were collected at birth, but not after birth; thus exposures occurring during childhood outside of the immediate neonatal period could not be directly assessed. Finally, some of the individual study sites have relatively small sample sizes and thus do not have the statistical power to detect some associations. Despite these limitations, the Collaborative Perinatal Study has been invaluable in terms of providing information relating to perinatal exposures and the subsequent development of adult psychiatric disorders.

Population-Based Studies

Studies based on epidemiological data obtained from defined populations have also been important in terms of identifying perinatal exposures and subsequent risks of adult psychiatric disorders. These studies have the advantage of allowing for the study of a large number of individuals in different populations over an extended period of time. They also allow for the comparison of the rates of disease across populations. Population-based studies have some limitations since they rely on data and information that are often collected for other purposes. These studies thus vary in terms of case definition and the completeness of their case finding. Also, the study of population-based cohorts generally does not allow for the analysis of blood or other biological samples. Despite these limitations, population-based studies have provided a great deal of information regarding large-scale exposures. Perinatal risk factors identified by population-based studies include season of birth [23], location of birth [12], household crowding [11], and birth during time of famine [24]. Population-based studies have also documented increased rates of schizophrenia occurring in individuals born during periods when there are increased population rates of stillbirths and specific infectious diseases [25]. It is of note that these studies generally do not distinguish between exposures during pregnancy and early life since the indicated environmental factors could be operant during both periods of development.

Register-Based Studies

The value of population-based studies has recently been enhanced by the availability of population studies based on the analysis of comprehensive medical registers. In these studies, all of the cases of schizophrenia in a population are identified from case registers and then linked to medical information relating to the mother's pregnancy and to the childhood health of the person who develops schizophrenia in later life [26]. The ability of register-based systems to provide information on environmental exposures has been enhanced by the recent finding that blood samples obtained for the diagnosis of neonatal metabolic diseases such as phenylketonuria can be retrieved and tested for antibodies to infectious agents, thus providing direct evidence of infectious exposures to the mother prior to the birth of the child [27]. While of great potential value, register-based studies are limited to societies where health registers are maintained and procedures are in place to access the health information. Up to this point in time, this situation has only existed in European countries and has largely been employed in Denmark, Sweden, and Finland, although register-based studies are being set up in other areas of the world with comprehensive systems of health care.

There are, however, also limitations inherent in register-based studies. One is that the diagnoses are largely based on records and are usually not confirmed by direct patient interviews or examinations. Secondly, register-based studies can only study participants in the nation's health care system. While most of the countries with register-based studies have universal health care systems, inclusion of immigrant and emigrant populations is often incomplete, allowing for some bias in terms of case finding. Despite these limitations, register-based studies have been extremely valuable in providing updated information concerning the relationship between a range of environmental exposures and psychiatric disorders. Examples of risk factors identified from register-based studies include associations between schizophrenia and urban birth, exposure to complications of pregnancy such as preeclampsia [28], exposure to mothers who have serological evidence of *Toxoplasma* infection [27], and central nervous system infections in childhood [29]. The continued application of the current register-based systems and the development of additional ones will undoubtedly supply a great deal of additional information relating to exposures during pregnancy and early childhood and psychiatric diseases in later life.

Perinatal Infectious Disease Exposures Associated with Schizophrenia

Prospective cohort, population-, and register-based studies have been crucial in terms of establishing a link between exposure to infectious agents during prenatal development and early life, and appearance of the symptoms of schizophrenia during adolescence or adulthood (Table 8.1). Exposures that have been identified include specific infectious agents such as *T. gondii*, rubella virus, poliovirus, and herpes simplex virus type 2, as well as infectious agents causing bacterial and viral meningitis. It is evident from the data presented in the table that there are several different types of infectious agents which are associated with an increased risk of schizophrenia following perinatal infection or infection during early life. Also, some studies have failed to find an association between these infectious agents and increased risk within certain populations, suggesting that there are differences among populations in terms of risk and susceptibility. It is also likely that some negative studies did not have a sufficient sample size to detect small effects within their study population. These findings underscore the importance of performing

Table 8.2 Mechanisms by which increased rates of maternal infection may lead to increased risk of schizophrenia in the offspring

Infectious disease mechanisms
Transmission of an infectious agent from the mother to the fetus across the placenta
Neonatal acquisition of an infectious agent during the birth process
Postnatal transmission of an infectious agent from the mother to the child
Exposure to common infectious agents within the family environment
Immunological mechanisms
Fetal exposure to cytokines and other immune modulators
Fetal exposure to maternal IgG antibodies which are transmitted to the fetus across the placenta and which cross-react with brain proteins
Genetic mechanisms
Common genes shared by mother and child associated with increased susceptibility to infection
Epigenetic effects of infection on fetal DNA

studies employing large sample sizes and diverse populations in order to obtain a comprehensive picture of exposure and disease risk.

The fact that different infectious agents are associated with increased risk of schizophrenia also suggests that different pathogenic mechanisms may be operative in the expression of this risk. Possible mechanisms of this association are depicted in Table 8.2. These include infectious disease mechanisms such as direct infection of the fetus or newborn, postnatal infection, infection from a common environmental source, and immunological mechanisms such as the effects of antibodies, cytokines, or other immune mediators [8, 30]. This possibility is supported by a number of animal models of infection and brain development [31–33]. In all of these scenarios, maternal and infant genotypes are also likely to play an important role in terms of the immune response to infection, as well as in terms of susceptibility of the developing brain to the effects of infection and immune activation [34, 35].

T. gondii and Risk of Schizophrenia

The T. gondii Organism

Among all of the factors which have been associated with increased risk of schizophrenia following perinatal exposure, *T. gondii* is selected for detailed review and analysis. The reasons for this selection are as follows: (1) *Toxoplasma* antibodies during pregnancy have been associated with increased risk of schizophrenia in several independent studies [27, 36, 37]; (2) Serological evidence of infection with *Toxoplasma* has also been associated with increased risk of psychiatric disorders in adults [38, 39]; (3) The biology of *Toxoplasma* is consistent with its ability to cause psychiatric disorders [40]; (4) The availability of new medications for the treatment of *Toxoplasma* infections holds the possibility of new interventions for the prevention and treatment of this disorder [41, 42].

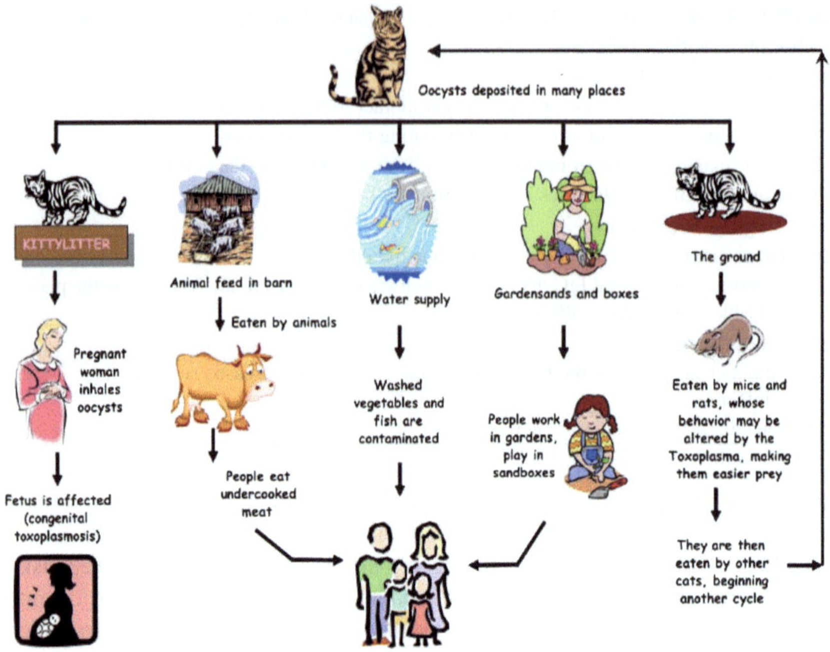

Fig. 8.1 Life cycle of *Toxoplasma gondii* (reprinted with permission from Yolken RH, Torrey EF (2008) Are some cases of psychosis caused by microbial agents? A review of the evidence. Mol Psychiatry 13(5):470–479)

Toxoplasma is an apicomplexan parasite with worldwide distribution. The life cycle of *T. gondii* is outlined in the Fig. 8.1. The primary hosts for *Toxoplasma* are cats and other felids. *Toxoplasma* can undergo its complete life cycle within the cat and can be transmitted from cat to cat both horizontally, through the excretion of oocysts in cat feces and ingestion by uninfected cats, or vertically, by transmission from mother to fetus [43, 44]. As in the case of many primary hosts, *Toxoplasma* undergoes high levels of replication in the cat but generally does not cause symptoms in otherwise healthy cats outside of the neonatal period [45]. Humans and other nonfelines can also become infected by the ingestion of oocysts derived from cat feces. This can occur by inhalation or ingestion of oocysts deposited in soil or by drinking water contaminated with oocysts [46]. Carnivorous or omnivorous animals such as humans can also become infected by the ingestion of tissue cysts formed in the meat of infected animals. The presence of multiple modes of transmission to secondary hosts such as humans insures a high level of prevalence in many populations.

Unlike primary hosts, secondary hosts such as humans have a number of symptoms related to *Toxoplasma* infection. These are most apparent in immune compromised hosts such as individuals with acquired or inherited immunodeficiency disorders, as well as in neonates whose mothers underwent primary infections [47, 48]. Such individuals can suffer from prolonged fever, hematologic abnormalities, severe eye disease, and diffuse encephalitis. However, recent studies of outbreaks indicate that many immune sufficient adults also have a range of symptoms including headache, retinitis, and lymphadenopathy in response to *Toxoplasma* infection, particularly following the ingestion of oocysts from contaminated water [49].

It is of note that humans and other secondary hosts are "dead ends" for *Toxoplasma* in that the organism cannot undergo sexual reproduction and complete its life cycle in these carriers. Recent experiments in rodent models have indicated that *Toxoplasma* organisms apparently attempt to overcome this problem by blunting the fear response and thus altering host behavior in a manner that encourages consumption of a *Toxoplasma*-infected rodent by a cat, and thus allows for the completion of the *Toxoplasma* life cycle, to the detriment of the rodent [50, 51]. This effect is likely to be vestigial in humans since modern humans are seldom consumed by cats. However, such predation may have been common in early human civilizations [52]. Furthermore, several studies have shown that serological evidence of *Toxoplasma* infection may be associated with altered fear response in otherwise healthy humans; instead of increased predation by cats, this blunted fear response may be manifested by other altered behaviors such as an increased rate of automobile accidents [53, 54]. The mechanisms by which *Toxoplasma* infection can lead to altered behavior is a subject of ongoing investigation. One recent discovery of note is that *Toxoplasma* organisms encode several amino acid hydroxylases that are capable of altering dopamine levels in the brain [55]. This finding, if confirmed, would be of great interest since in addition to being a critical component of the fear response, altered dopamine levels are a central feature of the pathobiology of schizophrenia [56], and most medications for schizophrenia are directed at the modulation of dopamine binding to cellular receptors [101].

Maternal Antibodies During Pregnancy

Increased levels of antibodies to *T. gondii* during pregnancy have been associated with an increased rate of schizophrenia in adult life in two independent studies. One of these studies was based on a prospective birth cohort in the United States [37], another on a population-based registry in Sweden [27]. The methodology and results of these studies are shown in Table 8.3. A third study, also based on a prospective cohort study in the United States, found a risk associated with increased levels of maternal antibodies, but the risk was largely associated with antibodies to only one of the *Toxoplasma* serotypes [57]. Interestingly, this serotype, known as type 1, has been shown to be associated with increased virulence in humans and animals [58, 59].

The biological mechanisms which define the risk of being born to a *Toxoplasma* seropositive mother have not yet been defined but are likely to encompass some or all of the mechanisms depicted in Table 8.2. One possibility, for example, is that fetuses undergo a direct *Toxoplasma* infection from the mother as occurs in

Table 8.3 Serology-based studies of early *Toxoplasma gondii* infections and schizophrenia risk

	Brown et al. [37]	Mortensen et al. [27]
Study population	63 Cases of schizophrenia and schizophrenia spectrum disorders	71 Cases of schizophrenia
		342 Cases with other psychoses or affective disorders
	In most cases, diagnoses based on research interview	Clinical diagnoses made by treating senior psychiatrist
	123 Controls	684 Controls
Setting	1959–1967 Birth cohort from the Kaiser Permanente Health plan, California, including treatment records and maternal sera drawn during pregnancy	All children born in Denmark 1981 or later, including treatment records for cases, controls, and their parents and sibs; neonatal blood samples from the national PKU-screening program
Design	Nested case-control study	Nested case-control study
Exposures measured	*Toxoplasma gondii* IgG and IgM titres from the maternal sera	*Toxoplasma gondii* IgG (maternal origin) and IgM titres from the child's neonatal blood samples
Confounders examined or controlled for	Age, gender, maternal age, maternal SES, ethnicity and education, timing of maternal sample	Age, gender, maternal age, psychiatric treatment history of parents and sibs, year, and place of birth
Association with elevated anti-*Toxoplasma gondii* IgG	OR 2.61 (1.00–6.82)	OR 1.79 (1.01–3.15)
		No significant association with schizophrenia-related disorders or affective disorders

Reprinted from [36]
Note: *PKU* phenylketonuria, *SES* socioeconomic status

congenital Toxoplasmosis. If so, the infections associated with increased risk of schizophrenia are largely subclinical since the infants who develop schizophrenia in later life do not have any of the manifestations of congenital Toxoplasmosis, such as microcephaly or cerebral calcifications. Also, the measurement of IgM antibodies in blood samples from the neonate does not indicate the presence of congenital infection [27]. It is thus likely that indirect mechanisms are governing this association. One possibility is that the mother undergoes *Toxoplasma* infection and, while she does not directly transmit the *Toxoplasma* organism to the fetus, she does transmit cytokines, antibodies, and other inflammatory markers which can alter brain development and lead to psychiatric disorders in later life [30, 60, 61]. This scenario is supported by studies in animal models, as well as a limited number of studies indicating that increased levels of maternal cytokines and neurotrophic factors may be associated with the development of schizophrenia in the offspring [62, 63].

Another possible mechanism by which maternal antibodies to an infectious agent may be associated with the development of schizophrenia in later life involves shared environmental risk. In the case of *Toxoplasma*, for example, it is possible that the mother and infant may both be exposed to *Toxoplasma* organisms through contact with a cat or the ingestion of contaminated food or water. The scenario of shared cat contact being a risk factor is supported by studies indicating that early life exposure to a pet cat, but not a pet dog, is associated with an increased risk of schizophrenia in adulthood [17]. It should be noted that, while for other agents such as the human herpesviruses, postnatal transmission from mother to child is a possible route of childhood exposure to an infectious agent, this is not likely to happen in the case of *Toxoplasma* since humans, as secondary hosts, do not shed oocysts and thus cannot transmit infection by this route. Knowledge of the mechanisms by which maternal exposure to infectious agents can lead to psychiatric disorders in the offspring is hampered by the lack of prospective cohort studies in which blood and other biological samples are collected from infants and children outside of the immediate neonatal period. Prospective cohort studies such as the National Children's Study described above [20] should provide the framework necessary to address these issues and to guide the development of programs to interrupt the cycle of disease transmission by whatever routes are found to be important within a given environment.

Toxoplasma Antibodies in Adults

Increased levels of *Toxoplasma* antibodies have also been associated with an increased risk of schizophrenia in adults. A recent meta-analysis based on more than 40 studies from around the world found that *Toxoplasma* antibodies were associated with schizophrenia with a pooled odds ratio of 2.73 (95% confidence interval, 2.10–3.60, $p<0.00001$) [38]. This odds ratio, while modest in terms of many infectious diseases, is quite high compared to other factors that have been linked to schizophrenia, such as specific genes which are generally associated with an odds ratio of <1.5 [64, 65].

Increased levels of IgG antibodies have also been found in cerebrospinal fluid samples obtained from individuals with a recent onset of schizophrenia, suggesting that the replication of *Toxoplasma* occurred at some point in time in the central nervous system of these individuals [66]. It should be noted that cross-sectional studies involving *Toxoplasma* antibodies and risk of schizophrenia do not document the timing of infection in comparison to disease onset. For example, it is possible that some of the risk is associated with primary infection around the time of disease onset. However, it is also possible that the increased levels of antibodies represent the effect of cerebral reactivation of *Toxoplasma* tissue cysts acquired from infection in early life. Finally, it is possible that some of the individuals at increased risk are reinfected with a different serotype of *Toxoplasma*, as has been shown to occur in animal models of *Toxoplasma* infection and in some cases of human perinatal infections [67, 68]. Prospective cohort studies will be required to address the question of timing of infection and to develop effective strategies for disease prevention.

Prevention and Treatment

One of the most important reasons for attempting to document the association between an infectious agent and a chronic disease such as schizophrenia is to develop strategies for the prevention and treatment of the infecting microbial agents. In order to be useful for the prevention or treatment of schizophrenia, an intervention would need to be relatively nontoxic and inexpensive since it would have to be given to a wide range of individuals, most of whom would also be taking other medications. The medication would also need to be able to pass through the blood–brain barrier if it were to be used for the treatment of infections within the central nervous system. In the case of *Toxoplasma*, therapeutic interventions that meet these criteria are limited since most of the available medications have significant levels of toxicity associated with their use [69]. However, recently a number of medications have been developed that have promise in terms of possessing both anti-*Toxoplasma* activity and low levels of toxicity for human cells. Many of these medications, such as the trioxane drugs derived from artemisinin, were originally developed for the treatment of malaria and are likely to owe their anti-*Toxoplasma* activity to shared pathways with this other apicomplexan organism [41, 70]. It is of note that some of these medications can be administered safely to pregnant women and thus have the potential of being used in an effort to prevent transmission of *Toxoplasma* from mother to infant [71]. Further development of safe and effective anti-*Toxoplasma* medications and their administration to at-risk individuals will be important in terms of establishing more of a link between *Toxoplasma* infection and psychiatric diseases, as well as providing new modalities for their prevention and treatment.

Unanswered Questions Relating to *Toxoplasma* and Schizophrenia

As discussed above, there is a growing body of data indicating a likely association between *Toxoplasma* and some cases of schizophrenia. However, there also are questions that remain in terms of the association between *Toxoplasma* infection and schizophrenia.

Individual Variation in Response to Toxoplasma Infection

Genetic

While epidemiological studies have documented a clear association between serological evidence of *Toxoplasma* infection and risk of schizophrenia, the risk of *Toxoplasma* infection is much more difficult to predict on an individual basis. This limitation is due to the fact that most individuals with serological evidence of *Toxoplasma* infection do not develop schizophrenia and do not have any other evidence of psychiatric abnormality. Reasons for the individual variation to *Toxoplasma* exposure are not known with certainty but might be related to variation in the pathogenicity of the infecting organism, individual differences in host resistance factors, or a combination of host and organism-related factors. In terms of the organism, possible variables which could affect virulence include the genetic makeup of the organism as well as the infectious dose, and the life form of the infecting organism. Thus, there might be a differential effect associated with the ingestion of tachyzoites, as might occur with the ingestion of water or soil contaminated with oocysts from cat feces, as compared with ingestion of bradyzoites from cysts found in infected meat.

In terms of the genotype of the organism, as noted above, there are at least three types of *Toxoplasma* that have been identified, and these vary in pathogenicity in both animal models and naturally occurring human infection [58]. Recently, assays have been devised which allow for the characterization of immune response in terms of these serotypes [72]. Preliminary studies using these assays indicate that serotype 1 is more likely to be associated with schizophrenia both in terms of mothers of infants who develop schizophrenia as well as adult individuals with this disorder. These assays should be applied more widely in order to better elucidate the pathogenic potential of the different serotypes of *Toxoplasma* in additional populations. The further availability of assays for the measurement of antibodies to the tachyzoite and bradyzoite forms of the organism would represent a major step forward in terms of elucidating the relationship between infection with the different life forms of the *Toxoplasma* organism and clinical outcome.

Immunogenetic

Differential response to *Toxoplasma* infection is also undoubtedly defined by host genes and other factors, which define the host immune response. Host genes which control the immune response to *Toxoplasma* infection have been well described in animal models. Most of these genes are related to the major histocompatibility complex [73], toll-like receptors [74], and other components of the immune response to infection. Genetic markers have also been identified that are associated with the course of congenital *Toxoplasma* infection in human neonates [75]. However, such genes have been less well defined in terms of natural infection of nonimmune compromised human children and adults [76]. It can be predicted that, in terms of susceptibility to schizophrenia, genes related to the development of schizophrenia in *Toxoplasma* infected individuals will involve both the immune system as well as the development of the brain and central nervous system. The identification of genetic factors that define the risk of schizophrenia in *Toxoplasma*-infected individuals and individuals born to *Toxoplasma*-infected mothers would be an important step in terms of defining the pathobiology of this link and in terms of identifying high-risk individuals who would be candidates for therapeutic interventions.

Geographic Differences in Antibody Prevalence

There are also unanswered questions relating to geographic differences in the prevalence of *Toxoplasma* antibodies and corresponding rates of schizophrenia within a population. For example, while the prevalence of *Toxoplasma* in adults living in the United States and Scandinavia ranges from 10 to 15%, the prevalence is higher in many other areas of the world. Thus, the prevalence of *Toxoplasma* antibodies in factory workers between 18 and 45 years of age in Addis Ababa, Ethiopia is 80% [77], and the *Toxoplasma* prevalence in pregnant women living in northeastern Brazil is 66% [78]. A high prevalence of antibodies to *Toxoplasma* has also been reported in both native born and immigrant women in Paris, France as recently as the 1980s [79]. The reasons for the high levels of antibodies in these populations is probably related to increased rates of oocyst ingestion from contaminated water and tissue cyst ingestion from undercooked infected meat. In any case, areas of the world with high rates of prevalence of antibodies to *Toxoplasma* are not known to be associated with increased rates of schizophrenia or other psychiatric disorders compared with areas with lower rates of exposure. The reasons for this discrepancy are not known with certainty but are likely to be related to the same factors that define individual response to infection, namely, the dose and timing of infection, the serotype and biological form of the infecting organism, and the genetic makeup of the host population. It is also possible that there are geographical differences in terms of the identification and diagnosis of individuals with schizophrenia.

The rate of *Toxoplasma* infection in many areas of the developed world is decreasing, probably as a result of improved methods for water purification and meat preparation. For example, the overall *Toxoplasma* prevalence measured in teenagers and young adults in the United States in the National Health and Nutrition Examination Survey (NHANES) study declined from 14.1% in 1988–1994 to 9.0% in 1999–2004 [80]. Similar declines have been noted in European populations as well [81]. However, these decreases have not led to an apparent decline in the rate of schizophrenia in the general population. In addition to the factors mentioned above, it is of note that infants born to mothers in the era of declining *Toxoplasma* prevalence will not reach the age of risk for schizophrenia until around the year

2020, so the continued study of the prevalence of *Toxoplasma* and schizophrenia in these populations may lead to new insights in terms of the timing of *Toxoplasma* infection and the risk of schizophrenia.

Despite the increased prevalence of *Toxoplasma* in individuals with schizophrenia and their mothers, most individuals with schizophrenia who live in low prevalence areas such as the United States do not have serological evidence of *Toxoplasma* infection. The most likely reason for this is that schizophrenia is a multifactorial disease and *Toxoplasma* does not play a direct role in the etiology of most of the cases within the population. However, it is also possible that *Toxoplasma* seropositivity does not persist for extended periods of time and thus that exposure during the perinatal period or in early childhood may not be manifested by antibodies measured later in life [82]. In addition, drugs used for the treatment of schizophrenia may have anti-*Toxoplasma* activity and thus decrease the prevalence of antibodies to *Toxoplasma* in treated individuals [41]. As discussed above, the best approach to address this question is the longitudinal study of *Toxoplasma* infection and schizophrenia risk throughout childhood, adolescence, and early adulthood. The recent finding of *Toxoplasma* encoded tyrosine and phenylalanine hydroxylases, which have the potential to alter levels of dopamine within the central nervous system [40], as discussed above, also suggests that other protozoa antigenically distinct from *Toxoplasma* may also play a role in the etiology of schizophrenia. For example, *Neospora caninum* is an apicomplexan parasite that is genetically related to *Toxoplasma* but does not demonstrate cross-reactivity at the level of serological assays [83]. While the protozoan does contain a phenylalanine hydroxylase which is highly homologous with that of *T. gondii*, human infection with *Neospora* species in most populations appear to be limited [84]. On the other hand, human infection with strains of the *Leishmania* organism is widespread in many areas of the world, and *Leishmania* encodes hydroxylases that are highly homologous to those of *T. gondii*. Serological studies of *Leishmania* species should be performed to explore the possible association of this parasite with schizophrenia and other psychiatric disorders. Additional protozoa should also be investigated for the presence of potentially neuroactive enzymes and their prevalence should be determined in human populations with psychiatric disorders and controls.

Conclusion

There is a growing body of evidence indicating that exposure to infectious agents during pregnancy and early life is a risk factor for the development of schizophrenia during adulthood. The evidence is particularly strong for the protozoan *T. gondii*; other organisms that share pathogenic and biological features with *Toxoplasma* are likely to be involved as well. The mechanisms for this association are likely to be varied and include direct infection, exposure to common environmental factors, and the maternal–fetal transfer of inflammatory mediators. Host genetic factors are likely to play a major role in terms of individual susceptibility to the different infectious agents. An improved understanding of the role of perinatally transmitted infectious agents and host factors in the etiopathogenesis of schizophrenia and related disorders will lead to new methods for the prevention and treatment of these devastating disorders.

References

1. Tienari P, Wynne LC, Läksy K et al (2003) Genetic boundaries of the schizophrenia spectrum: evidence from the Finnish Adoptive Family Study of Schizophrenia. Am J Psychiatry 160(9):1587–1594.
2. Greenwood TA, Braff DL, Light GA et al (2007) Initial heritability analyses of endophenotypic measures for schizophrenia: the consortium on the genetics of schizophrenia. Arch Gen Psychiatry 64(11):1242–1250.
3. Sanders AR, Duan J, Levinson DF et al (2008) No significant association of 14 candidate genes with schizophrenia in a large European ancestry sample: implications for psychiatric genetics. Am J Psychiatry 165(4):497–509.
4. Kirov G, Grozeva D, Norton N et al (2009) Support for the involvement of large copy number variants in the pathogenesis of schizophrenia. Hum Mol Genet 18(8):1497–1503.
5. Need AC, Ge D, Weale ME et al (2009) A genome-wide investigation of SNPs and CNVs in schizophrenia. PLoS Genet 5(2):e1000373, Epub 2009 Feb 6.
6. Allen NC, Bagade S, McQueen MB et al (2008) Systematic meta-analyses and field synopsis of genetic association studies in schizophrenia: the SzGene database. Nat Genet 40(7):827–834.
7. Byrne M, Agerbo E, Bennedsen B, Eaton WW, Mortensen PB (2007) Obstetric conditions and risk of first admission with schizophrenia: a Danish national register based study. Schizophr Res 97(1–3):51–59.
8. Schlotz W, Phillips DI (2009) Fetal origins of mental health: evidence and mechanisms. Brain Behav Immun 23(7):905–916.
9. Kinney DK, Teixeira P, Hsu D, Napoleon SC, Crowley DJ, Miller A, Hyman W, Huang E (2009) Relation of schizophrenia prevalence to latitude, climate, fish consumption, infant mortality, and skin color: a role for prenatal vitamin D deficiency and infections? Schizophr Bull 35(3):582–595.
10. Torrey EF, Torrey BB, Peterson MR (1977) Seasonality of schizophrenic births in the United States. Arch Gen Psychiatry 34(9):1065–1070.
11. Agerbo E, Torrey Ef, Mortensen PB (2001) Household crowding in early adulthood and schizophrenia are unrelated in Denmark: a nested case-control study. Schizophr Res 34(9):1065–1070.
12. Pedersen CB, Mortensen PB (2001) Family history, place and season of birth as risk factors for schizophrenia in Denmark: a replication and reanalysis. Br J Psychiatry 17:46–52.
13. Cantor-Graae E, Pedersen CB (2007) Risk of schizophrenia in second-generation immigrants: a Danish population-based cohort study. Psychol Med 37(4):485–494.
14. Segal S, Hill AV (2003) Genetic susceptibility to infectious disease. Trends Microbiol 11(9):445–448.
15. Lewis SW, Murray RM (1987) Obstetric complications, neurodevelopmental deviance, and risk of schizophrenia. J Psychiatr Res 21(4):413–421.
16. Wright P, Takei N, Rifkin L, Murray RM (1997) Maternal influenza, obstetric complications, and schizophrenia. Am J Psychiatry 154(2):292–293.
17. Torrey EF, Rawlings R, Yolken RH (2000) The antecedents of psychosis: a case-control study of selected risk factors. Schizophr Res 46(1):17–23.

18. Buka SL, Goldstein JM, Spartos E, Tsuang MT (2004) The retrospective measurement of prenatal and perinatal events: accuracy of maternal recall. Schizophr Res 712(2–3):417–426.
19. Buka SL, Goldstein JM, Seldman LI, Tsuang MT (2000) Maternal recall of pregnancy history: accuracy and bias in schizophrenia research. Schizophr Bull 26(2):335–350.
20. Landrigan PJ, Trasande L, Thorpe LE et al (2006) The National Children's Study: a 21-year prospective study of 100,000 American children. Pediatrics 118(5):2179–2186.
21. Horwood J, Salvi G, Thomas K, Duffy L, Gunnell D, Hollis C, Lewis G, Menezes P, Thompson A, Wolke D, Zammit S, Harrison G (2008) IQ and non-clinical psychotic symptoms in 12-year-olds: results from the ALSPAC birth cohort. Br J Psychiatry 193(3):185–191.
22. Shiono PH, Klebanoff MA, Berendes HW (1986) Congenital malformations and maternal smoking during pregnancy. Teratology 34(1):65–71.
23. Davies G, Welham J, Chant D, Torrey EF, McGrath J (2003) A systematic review and meta-analysis of Northern Hemisphere season of birth studies in schizophrenia. Schizophr Bull 29(3):587–593.
24. Xu MQ, Sun WS, Liu BX et al (2009) Prenatal malnutrition and adult schizophrenia: further evidence from the 1959–1961 Chinese famine. Schizophr Bull 35:568–576.
25. Torrey EF, Rawlings R, Waldman IN (1988) Schizophrenic births and viral diseases in two states. Schizophr Res 1(1):73–77.
26. Löffler W, Häfner H, Fätkenheuer B et al (1994) Validation of Danish case register diagnosis for schizophrenia. Acta Psychiatr Scand 90(3):196–203.
27. Mortensen PB, Nørgaard-Pedersen B, Waltoft BL et al (2007) *Toxoplasma gondii* as a risk factor for early-onset schizophrenia: analysis of filter paper blood samples obtained at birth. Biol Psychiatry 61(5):688–693.
28. Dalman C, Allebeck P, Culberg J, Grunewald C, Köster M (1999) Obstetric complications and the risk of schizophrenia: a longitudinal study of a national birth cohort. Arch Gen Psychiatry 56(3):234–240.
29. Dalman C, Allebeck P, Gennell D et al (2008) Infections in the CNS during childhood and the risk of subsequent psychotic illness: a cohort study of more than one million Swedish subjects. Am J Pshchiatry 165(1):59–65.
30. Nahmias AJ, Nahmias SB, Danielsson D (2006) The possible role of transplacentally-acquired antibodies to infectious agents, with molecular mimicry to nervous system sialic acid epitopes as causes of neuromental disorders: prevention and vaccine implications. Clin Dev Immunol 13(2–4):167–183.
31. Winter C, Djodari-Irani A, Sohr R et al (2008) Prenatal immune activation leads to multiple changes in basal neurotransmitter levels in the adult brain: implications for brain disorders of neurodevelopmental origin such as schizophrenia. Int J Neuropsychopharmacol 28:1–12.
32. Shi L, Smith SE, Malkova N et al (2009) Activation of the maternal immune system alters cerebellar development in the offspring. Brain Behav Immun 23(1):116–123.
33. Fatemi SH, Reutiman TJ, Folsom TD et al (2008) Maternal infection leads to abnormal gene regulation and brain atrophy in mouse offspring: implications for genesis of neurodevelopmental disorders. Schizophr Res 99(1–2):56–70.
34. Mittal VA, Ellman LM, Cannon TD (2008) Gene-environment interaction and covariation in schizophrenia: the role of obstetric complications. Schizophr Bull 34(6):1083–1094.
35. Kim JJ, Shirts BH, Dayal M et al (2007) Are exposure to cytomegalovirus and genetic variation on chromosome 6p joint risk factors for schizophrenia? Ann Med 39(2):145–153.
36. Mortensen PB, Nørgaard-Pederson B, Waltoft BL et al (2007) Early infections of *Toxoplasma gondii* and the later development of schizophrenia. Schizophr Bull 33(3):741–744.
37. Brown AS, Schaefer CA, Quesenberry CP et al (2005) Maternal exposure to toxoplasmosis and risk of schizophrenia in adult offspring. Am J Psychiatry 162(4):767–773.
38. Torrey EF, Bartko JJ, Lun ZR, Yolken RH (2007) Antibodies to *Toxoplasma gondii* in patients with schizophrenia: a meta-analysis. Schizophr Bull 33(3):729–736.

39. Zhu S, Guo MF, Feng QC, Fan JM, Zhang LX (2007) Epidemiological evidences from China assume that psychiatric-related diseases may be associated with *Toxoplasma gondii* infection. Neuro Endocrinol Lett 28(2):115–120.
40. Henriquez SA, Brett R, Alexander J, Pratt J, Roberts CW (2009) Neuropsychiatric disease and *Toxoplasma gondii* infection. Neuroimmunomodulation 16(2):122–133.
41. D'Angelo JG, Bordon C, Posner GH, Yolken RH, Jones-Brando L (2009) Artemisinin derivatives inhibit *Toxoplasma gondii* in vitro at multiple steps in the lytic cycle. J Antimicrob Chemother 63(1):146–150.
42. Krivogorsky B, Grundt P, Yolken R, Jones-Brando L (2008) Inhibition of *Toxoplasma gondii* by indirubin and tryptanthrin analogs. Antimicrob Agents Chemother 52(5):4466–4469.
43. Hill D, Dubey JP (2002) *Toxoplasma gondii*: transmission diagnosis and prevention. Clin Microbiol Infect 8(10):634–640.
44. Dubey JP (1998) Advances in the life cycle of *Toxoplasma gondii*. Int J parasitol 28(7):1019–1024.
45. Powell CC, Lappin MR (2001) Clinical ocular toxoplasmosis in neonatal kittens. Vet Ophthalmol 4(2):87–92.
46. Dubey JP (2004) Toxoplasmosis – a waterborne zoonosis. Vet Parasitol 126(1–2):57–72.
47. Montoya JG, Liesenfeld O (2004) Toxoplasmosis. Lancet 363(9425):1965–1976.
48. Boyer KM, Holfels E, Roizen N et al (2005) Risk factors for *Toxoplasma gondii* infection in mothers of infants with congenital toxoplasmosis: implications for prenatal management and screening. Am J Obstet Gynecol 192(2):564–571.
49. Bowie WR, King AS, Werker DH et al (1997) Outbreak of toxoplasmosis associated with municipal drinking water. The BC Toxoplasma Investigation Team. Lancet 350(9072):173–177.
50. Lamberton PH, Donnelly CA, Webster JP (2008) Specificity of the *Toxoplasma gondii*-altered behavior to definitive versus non-definitive host predation risk. Parasitology 135(10):1143–1150.
51. Webster JP (2007) The effect of *Toxoplasma gondii* on animal behavior: playing cat and mouse. Schizophr Bull 33(3):752–756.
52. Webster JP, Berdoy M (1997) Characteristics and evolution of parasite-altered behavior. In: Holland CV (ed) Perspectives in zoonoses. Royal Irish Academic Press, Dublin, pp 115–121.
53. Flegr J, Havlícek J, Kodym P et al (2002) Increased risk of traffic accidents in subjects with latent toxoplasmosis: a retrospective case-control study. BMC Infect Dis 2:11.
54. Yereli K, Balcioğlu IC, Ozbilgin A (2006) Is *Toxoplasma gondii* a potential risk for traffic accidents in Turkey? Forensic Sci Int 163(1–2):34–37.
55. Gaskell EA, Smith JE, Pinney JW et al (2009) A unique dual activity amino acid hydroxylase in *Toxoplasma gondii*. PLoS One 4(3):e4801, Epub 2009 Mar 11.
56. Meyer U, Feldon J (2009) Prenatal exposure to infection: a primary mechanism for abnormal dopaminergic development in schizophrenia. Psychopharmacology (Berl) 206(4):587–602.
57. Xiao J-C, Buka S, Cannon T et al (2009) Serological pattern consistent with infection with type I *Toxoplasma gondii* in mothers and risk of psychosis among adult offspring. Microbes Infect 206(4):587–602.
58. Morisset S, Peyron F, Lobry JR et al (2008) Serotyping of *Toxoplasma gondii*: striking homogeneous pattern between symptomatic and asymptomatic infections within Europe and South America. Microbes Infect 10(7):742–747.
59. Nguyen TD, Bigaignon G, Markine-Goriaynoff D et al (2003) Virulent *Toxoplasma gondii* strain RH promotes T-cell independent overproduction of proinflammatory cytokines IL12 and gamma-interferon. J Med Microbiol 52(Pt 10):869–876.
60. Nawa H, Takei N (2006) Recent progress in animal modeling of immune inflammatory processes in schizophrenia: implication of specific cytokines. Neurosci Res 56(1):2–13.
61. Tamer GS, Dundar D, Yalug I et al (2008) The schizophrenia and *Toxoplasma gondii* connection: infectious, immune or both? Adv Ther 25(7):703–709.
62. Buka SL, Tsuang MT, Torrey EF et al (2001) Maternal cytokine levels during pregnancy and adult psychosis. Brain Behav Immun 15(4):411–420.

63. Cannon TD, Yolken R, Buka S et al (2008) Decreased neurotrophic response to birth hypoxia in the etiology of schizophrenia. Biol Psychiatry 64(9):797–802.
64. Gilbody S, Lewis S, Lightfoot T (2007) Methylenetetrahydrofolate reductase (MTHFR) genetic polymorphisms and psychiatric disorders: a HuGE review. Am J Epidemiol 165(1):1–13.
65. Zintzaras E (2007) Brain-derived neurotropic factor gene polymorphisms and schizophrenia: a meta-analysis. Psychiatr Genet 17(2):69–75.
66. Leweke FM, Gerth CW, Koethe D et al (2004) Antibodies to infectious agents in individuals with recent onset schizophrenia. Eur Arch Psychiatry Clin Neurosci 254(1):4–8.
67. Elbez-Rubinstein A, Ajzenberg D, Dardé ML et al (2009) Congenital toxoplasmosis and reinfection during pregnancy: case report, strain characterization, experimental model of reinfection and review. J Infect Dis 199(2):280–285.
68. Carruthers VB, Suzuki Y (2007) Effects of *Toxoplasma gondii* infection on the brain. Schizophr Bull 33(3):745–751.
69. Georgiev VS (1994) Management of *Toxoplasma gondii* infection on the brain. Drugs 48(2):179–188.
70. Li Y, Wu YL (1998) How Chinese scientists discovered qinghaosu (artemisinin) and developed its derivatives? What are the future perspectives? Med Trop (Mars) 58(Suppl 3):9–12.
71. Milner DA Jr, Montgomery J, Seydel KB et al (2008) Severe malaria in children and pregnancy: an update and perspective. Trends parasitol 24(12):590–595.
72. Dardé ML (2008) *Toxoplasma gondii*, "new" genotypes and virulence. Parasite 15(3):366–371.
73. Lüder DG, Stanway RR, Chaussepied M et al (2009) Intracellular survival of apicomplexan parasites and host cell modification. Int J Parasitol 39(2):163–173.
74. Yarovinsky F (2008) Toll-like receptors and their role in host resistance to *Toxoplasma gondii*. Immunol Lett 119(1–2):17–21.
75. Jamieson SE, de Rubaix LA, Cortina-Boria M et al (2008) Genetic and epigenetic factors at COL2A1 and ABCA4 influence clinical outcome in congenital toxoplasmosis. PLos ONE 3(6):e2285.
76. Suzuki Y (2002) Host resistance in the brain against *Toxoplasma gondii*. J Infect Dis 185(Suppl 1):S58–S65.
77. Woldemichael T, Fontanet AL, Sahlu T et al (1998) Evaluation of the Eiken latex agglutination test for anti-Toxoplasma antibodies and seroprevalence of *Toxoplasma* infection among factory workers in Addis Ababa, Ethiopia. Trans R Soc Trop Med Hyg 92(4):401–403.
78. Barbosa IR, deCarvalho Xavier Holanda CM, de Andrade-Neto VF (2009) Toxoplasmosis screening and risk factors amongst pregnant females in Natal, northeastern Brazil. Trans R Soc Trop Med Hyg 103:377–382.
79. Jeannel D, Niel G, Costagliola D et al (1988) Epidemiology of toxoplasmosis among pregnant women in the Paris area. Int J Epidemiol 17(3):595–602.
80. Jones JL, Kruszon-Moran D, Sanders-Lewis K, Wilson M (2007) *Toxoplasma gondii* infection in the United States, 1999–2004, decline from the prior decade. Am J Trop Med Hyg 77(3):405–410.
81. Nash JQ, Chissel S, Jones J et al (2005) Risk factors for toxoplasmosis in pregnant women in Kent, United Kingdom. Epidemiol Infect 133(3):475–483.
82. Van Druten H, van Knapen F, Reinties A (1990) Epidemiologic implications of limited-duration seropositivity after *Toxoplasma* infection. Am J Epidemiol 132(1):169–180.
83. Tranas J, Heinzen RA, Weiss LM et al (1999) Serological evidence of human infection with the protozoan *Neospora caninum*. Clin Diagn Lab Immunol 6:765–767.
84. McCann CM, Vyse AJ, Salmon RL et al (2008) Lack of serologic evidence of *Neospora caninum* in humans, England. Emerg Infect Dis 14(6):978–980.
85. Buka SL, Tusang MT, Torrey EF et al (2001) Maternal infections and subsequent psychosis among offspring. Arch Gen Psychiatry 58(11):1032–1037.
86. Mortensen PB et al., submitted for publication.

87. Ellman LM, Yolken RH, Buka SL et al (2009) Cognitive functioning prior to the onset of psychosis: the role of fetal exposure to serologically determined influenza infection. Biol Psychiatry 65:1040–1047.
88. Brown AS, Begg MD, Gravenstein S et al (2004) Serologic evidence of prenatal influenza in the etiology of schizophrenia. Arch Gen Psychiatry 61:774–780.
89. Adams W, Kendell RE, Hare EH, Munk-Jørgensen P (1993) Epidemiological evidence that maternal influenza contributes to the aetiology of schizophrenia. An analysis of Scottish, English, and Danish data. Br J Psychiatry 163:522–534.
90. Westergaard T, Mortensen PB, Pedersen CB et al (1999) Exposure to prenatal and childhood infections and the risk of schizophrenia: suggestions from a study of sibship characteristics and influenza prevalence. Arch Gen Psychiatry 56(11):993–998.
91. Brown AS, Susser ES (2002) In utero infections and adult schizophrenia. Ment Retard Dev Disabil Res Rev 8(1):51–57.
92. Suvisaari J, Haukka J, Tanskanen A, Hovi T, Lönnqvist J (1999) Association between prenatal exposure to poliovirus infection and adult schizophrenia. Am J Psychiatry 156(7):1100–1102.
93. Cahill M, Chant D, Welham J, McGrath J (2002) No significant association between prenatal exposure poliovirus epidemics and psychosis. Aust N Z J Psychiatry 36(3):373–375.
94. O'Callaghan E, Sham PC, Takei N et al (1994) The relationship of schizophrenic births to 16 infectious diseases. Br J Psychiatry 165(3):353–356.
95. Sorensen HJ, Mortensen EL, Reinisch JM, Mednick SA (2009) Association between prenatal exposure to bacterial infection and risk of schizophrenia. Schizophr Bull 35(3):631–637, Epub 2008 Oct 1.
96. Babulas V, Factor-Litvak P, Goetz R et al (2006) Prenatal exposure to material genital and reproductive infections and adult schizophrenia. Am J Psychiatry 163(5):927–929.
97. Koponen H, Rantakallio P, Veijola J et al (2004) Childhood central nervous system infections and risk for schizophrenia. Eur Arch Psychiatry Clin Neurosci 254:9–13.
98. Suvisaari J, Mautemps N, Haukka J et al (2003) Childhood central nervous system viral infections and adult schizophrenia. Am J Psychiatry 160:1183–1185.
99. Opler MG, Buka SL, Groeger J et al (2008) Prenatal exposure to lead, delta-aminolevulinic acid, and schizophrenia: further evidence. Environ Health Perspect 116(11):1586–1590.
100. Kyle UG, Pichard C (2006) The Dutch famine of 1944–1945: a pathophysiological model of long-term consequences of wasting disease. Curr Opin Nutr Metab Care 9(4):388–394.
101. Jones-Brando L, Torrey EF, Yolken R (2003) Drugs used in the treatment of schizophrenia and bipolar disorder inhibit the replication of *Toxoplasma gondii*. Schizophr Res 62(3):237–244.

Chapter 9
Maternally Acting Alleles in Autism and Other Neurodevelopmental Disorders: The Role of HLA-DR4 Within the Major Histocompatibility Complex

William G. Johnson, Steven Buyske, Edward S. Stenroos, and George H. Lambert

Keywords Autism • Maternally acting gene alleles • Maternal effects • Pregnancy • Developmental genes • Maternal • Teratogenesis

Introduction

Autism is a common and devastating neurodevelopmental disorder that begins in early childhood and whose causes are believed to be the interactions of multiple genes with environmental factors. These genes are usually thought of as acting in the individual with autism. In this chapter, an additional category of genes is considered: maternal genes that act in the mothers (most likely) during pregnancy to contribute to the autism phenotype of their affected offspring.

Maternally acting alleles are novel contributors to neurodevelopmental disorders, including autism, and with time, more are being reported in this rapidly moving field. As with autism, neurodevelopmental disorders are considered to be complex disorders in which multiple genes contribute to the clinical phenotype and in which gene effects are modified by environmental factors. Nearly all of the genes that have been reported as contributing to neurodevelopmental disorders act in the affected individual, i.e., the child or adult with the neurodevelopmental disorder. However, some of the genes that are now being identified for these

W.G. Johnson (✉)
Department of Neurology, UMDNJ-Robert Wood Johnson Medical School, 671 Hoes Lane, Piscataway, NJ, 08854, USA
and
Center for Childhood Neurotoxicology & Exposure Assessment, UMDNJ-Robert Wood Johnson Medical School, 671 Hoes Lane, Piscataway, 08854, NJ, USA
e-mail: wjohnson@umdnj.edu

A.W. Zimmerman and S.L. Connors (eds.), *Maternal Influences on Fetal Neurodevelopment: Clinical and Research Aspects*,
DOI 10.1007/978-1-60327-921-5_9,

disorders are maternal genes that act in the mothers to contribute to the phenotype of their affected offspring. For these genes, the genetic "patient" is the *mother* of the individual with the neurodevelopmental disorder. The alleles of maternal genes that act in this way have been termed "teratogenic alleles" because their effects in some ways are similar to ingested maternal teratogens that affect fetal development [1]. However, unlike the effect of ingested maternal teratogens, the effect of maternally acting alleles ("teratogenic alleles") is of genetic, not environmental, origin and the term "maternally acting alleles" may be clearer. These maternal alleles act in the mother, most likely during pregnancy, by modifying development of the embryo or fetus, e.g., brain development in the affected children, although action in the ovum or on the ovum before conception is also possible. These maternal gene alleles may interact with fetal alleles and with environmental factors. Fetal genes may modify the severity of the phenotype in the fetus or determine which organ is affected. These considerations have been summarized as the Gene-Teratogen Model [1].

At present, at least 35 reports of these maternally acting gene alleles have been published (Table 9.1). Nearly all of these reports of maternally acting alleles involve neurodevelopmental disorders. Their number has more than doubled since the topic was first reviewed in 2003 [1]. It is possible that maternally acting alleles are a characteristic and even defining feature of neurodevelopmental disorders, but more work is needed to clarify their impact.

Early Examples of Maternally Acting Alleles

A few examples of what turned out to be maternally acting alleles were reported before the concept was well understood. Rh incompatibility was reported in 1939 and its genetic mechanism was clarified in 1940. Maternal phenylketonuria (PKU) was recognized as early as 1956 and the genetic mechanism of PKU itself was well understood still earlier to be an autosomal recessive inborn error of metabolism. These examples are important as a proof-of-principle for maternally acting alleles. This action was not documented by a genetic test, e.g., maternal transmission disequilibrium testing or the log–linear method. However, since the genetic mechanism of these disorders is now well known, no such documentation is required at the present time.

Rh Incompatibility

Rh incompatibility between mother and fetus was first described in 1939 by Levine and Stetson [34], who found a new antibody in the serum of a mother who had recently delivered a stillborn child and had subsequently had a reaction to an ABO-compatible transfusion from her husband. They correctly surmised that this

Table 9.1 Reports of maternally acting alleles

Maternal allele	Disease	Analysis by	References
Maternal TDT			
1. *MTR*2756G*	NTD/spina bifida	Mat TDT, log–linear	[2]
2. *MTRR*66G*	NTD/spina bifida	Mat TDT, log–linear	[2]
3. *GSTP1*val105*	Autism	Mat TDT	[3]
4. *HLA-DR4*	Autism	Mat TDT	[4]
Maternal TDT-equivalent			
5. Rh *d*	Erythroblastosis fetalis	Mat TDT equivalent	[5]
6. *PAH* mutations	Maternal PKU	Mat TDT equivalent	[6]
7. Rh *d*	Schizophrenia	Mat TDT equivalent, log–linear	[7, 8]
Case-parent log–linear analysis without maternal TDT or equivalent			
8. *MTHFD1*653Q*	Spina bifida	Log–linear	[9]
9. *CYP1A1*6235C*	Low birth weight	Log–linear	[10]
10. *NAT1* genotypes	NTD/spina bifida	Log–linear	[11]
11. *CCL-2*2518AA*	NTD/spina bifida	Log–linear	[12]
Regression analysis			
12. *APOE*E2*	Lower LDL-C, apoB, Higher HDL-C, apoa1 In newborns	Forward stepwise Regression analysis	[13]
13. *APOC3*S2*	Lower newborn LDL-C, apoB, HDL-C, apoA1	Forward stepwise Regression analysis	[13]
14. *LPL*S447X*	Lower newborn LDL-C, apoB, TG	Forward stepwise Regression analysis	[13]
15. *GSTP1*Val105, *Val114*	Asthma	Multiple linear regression	[14]
Case–control plus case TDT			
16. *MTHFR*677T*	Oral–facial clefting, cleft lip±cleft palate	Case–control, case TDT negative	[15]
17. *MTHFR 677T*	Congen. heart dis.	Case–control, case TDT negative	[16]
18. *MTHFR*677TT*	Down syndrome	Case–control, case TDT negative	[17]
19. *HLA-DR*4*	Autism	Case–control, case TDT negative	[18]

(continued)

Table 9.1 (continued)

Maternal allele	Disease	Analysis by	References
Case–control			
20. *C4B*0*	Autism	Case–control	[19]
21. *HLA-DR4*	Rheumatoid arthritis	Case–control	[20, 21]
22. *HLA-DR4*	Autism	Case–control	[22]
23. *GSTP1-1b*	Recurrent early pregnancy loss	Case–control	[23]
24. *MTHFR*1298C*	NTD/spina bifida	Case–control	[24]
25. *MTRR*66G *GG*	Down syndrome	Case–control	[25]
26. MTH*FR*677T, MTRR*66GG*	Down syndrome	Case–control	[25]
27. *GSTT1*0*	Oral–facial clefting	Case–control	[26]
28. *MTHFR*1298C*	NTD/spina bifida	Case–control	[27]
29. *GSTM1*0*	Recurrent pregnancy loss	Case–control	[28]
30. *DHFR*19bp del*	Spina bifida	Case–control	[29]
31. *CYP1A1*2A*	Recurrent pregnancy loss	Case–control	[30]
32. *DHFR*19bp del*	Preterm delivery	Case–control	[31]
33. *MTHFR*1298C*	Down syndrome	Case–control	[24]
34. *MTHFR*1298C*	Down syndrome	Case–control	[32]
35. *RFC1*A80G, GG*	Down syndrome	Case–control	[32]

Reports of maternally acting alleles are listed: (1) According to the method of analysis used and (2) according to the date of the report beginning with the earliest. The specific maternally acting allele is listed on the left, next the specific disease or disorder studied, next the study design and finally, the literature reference. The specific maternally acting allele is given using the nomenclature: gene symbol*allele designation. The names of the genes corresponding to the gene symbol are given in the text and correspond to the designation in Online Mendelian Inheritance in Man (OMIM). Gene symbols are given in upper case letters and italicized while the corresponding protein symbol is given in the same upper case letters but not italicized

Acronyms: Human genes are given in upper case italics, corresponding proteins in the same upper case letters that are not italicized

APOE apolipoprotein E, *APOC3* apolipoprotein C3, *C4B* complement component 4B gene, a blood group antigen, *CCL* monocyte chemoattractant protein gene, *CHD* congenital heart defect, *CYP1A1* cytochrome P450 1A1 gene, *DHFR* dihydrofolate reductase gene, *GSTM1* glutathione S-transferase M1 gene, *GSTP1* glutathione *S*-transferase P1 gene, *GSTT1* glutathione S-transferase T1 gene, *HLA* human leukocyte antigen system, *LPL* lipoprotein lipase, *OFC* oral–facial clefting, *Rh* Rhesus factor, a protein on the surface of human red blood cells that may be diminished or absent, *RhD,d* alleles of the *RhD* gene, *MTHFD1* methylenetetrahydrofolate dehydrogenase gene, *MTHFR* methylenetetrahydrofolate reductase gene, *MTR* methionine synthase gene, *MTRR* methionine synthase reductase gene, *NAT1* *N*-acetyltransferase gene, *NIMA* non-inherited maternal (gene) allele, *NIPA* non-inherited paternal (gene) allele, *NTD* neural tube defect, *PAH* phenylalanine hydroxylase gene, *RFC1* reduced folate carrier1 gene Table adapted and reprinted from [33] with permission of author and publisher

antibody was unrelated to any known blood group and resulted from immunization of the mother during pregnancy by an antigen of paternal origin carried by the fetus, and further, that the hemolytic transfusion reaction resulted from this maternal antibody reacting with the same antigen on the husband's transfused cells. This antibody was later shown to be the same as the anti-Rhesus (anti-Rh) prepared by Landsteiner and Wiener in 1940 [35].

The genetic architecture of Rh was presented by Fisher and Race [5] as three closely-linked loci producing C or c, D or d, and E or e. However, it turned out that RhD is a single gene while RhC,c and RhE,e are both part of a second homologous gene [36]. RhD hemolytic disease of the newborn (HDN, erythroblastosis fetalis) can cause mild or severe disease and is the most common form of severe HDN. RhE HDN is a mild disorder. Rh c HDN can produce mild or severe disease. RhC and e are rare causes of HDN. Combinations of antibodies, e.g., anti-Rh c and E, can produce severe disease. The molecular causes of RhD–HDN are quite complex. Depending upon which parts of the RhD gene are altered and in what fashion, RhD-alleles or D variants, e.g., weak D and partial D variant alleles, may result [37].

For Rh incompatibility to occur (Table 9.1, #5), the mother most commonly lacks the RhD antigen on the surface of her erythocytes and is thus Rh negative (RhD negative); that is, she is a homozygote for the Rh *d*-allele with the *dd* genotype. Alternatively, she may be a homozygote or a genetic compound for RhD alleles that have the same effect. Also the fetus must carry an Rh *D*-allele that can only be of paternal origin (i.e., father is RhD positive). During the pregnancy, the Rh *dd* mother is exposed to the RhD antigen produced by the fetus. Since the mother lacks this antigen, she makes antibodies to RhD antigen as the pregnancy progresses. During a subsequent pregnancy with an RhD-positive fetus, maternal antibodies are again produced but in greater amount and at an accelerated rate. With further RhD-positive fetuses, the mother mounts an immunological attack upon the fetus, who may develop erythroblastosis fetalis, which may lead to a neurodevelopmental disorder.

This developmental disorder consists of three clinical syndromes in the fetus and neonate: anemia of the newborn, neonatal jaundice that can lead to the severe fetal encephalopathy of kernicterus, and the generalized neonatal edema of hydrops fetalis with massive anasarca, pleural effusions and ascites. The maternally acting allele here is Rh *d*, for which the mother is a homozygote. Each of her parents carries an Rh *d*-allele and has transmitted an Rh *d*-allele to her. Thus, transmission disequilibrium is present. The same or equivalent disorder can result if the mother is a homozygote or a genetic compound for other RhD alleles that have the same effect or for certain RhC,c or RhE,e alleles. The mechanisms causing fetal damage are unclear, but cytokine abnormalities have been observed [38].

Maternal Phenylketonuria

Maternal PKU results from major intrauterine effects on fetuses of mothers with PKU [39] and was recognized as early as 1956 [6]. PKU itself is a recessive postnatal disorder. Untreated homozygous PKU mothers and fathers both have elevated blood phenylalanine. However, heterozygous offspring of untreated PKU mothers may develop maternal PKU, a disorder different from PKU, that has an abnormal developmental and neurodevelopmental phenotype [39–41]. Nearly all cases of PKU and hence maternal PKU are known to result from mutations in the phenylalanine hydroxylase (*PAH*) gene. Thus the mutations in the maternal *PAH* genes act on the fetus during pregnancy through elevation of maternal blood phenylalanine or other metabolite(s) in the untreated PKU mother (Table 9.1, #6).

Infants with PKU [42] are normal at birth and develop a progressive metabolic disorder of postnatal onset characterized by vomiting, eczema, mental retardation, and infantile spasms with a typical pattern called hypsarrhythmia on the electroencephalogram. In contrast, infants with maternal PKU [42] have a congenital nonprogressive disorder of fetal onset characterized by microcephaly, abnormal facies, mental retardation, congenital heart disease and prenatal and postnatal growth retardation. The effect of the maternally acting alleles in maternal PKU is not dependent upon the fetal genotype, although the fetus is an obligate heterozygote since the mother is a homozygote for *PAH* mutations and the father usually has a normal genotype. The maternally acting allele here is the *PAH* mutation for which the mother is a homozygote. The mother may also be a genetic compound with two different *PAH* mutations. In either case, each of her parents carries a *PAH* mutation and has transmitted it to her. Thus, transmission disequilibrium is present. Again, there is no need to demonstrate this at present since the disorder and its mechanism are well known.

More Recent Reports of Maternally Acting Alleles

Choice of a Study Design to Document Maternally Acting Alleles

The presence of a maternally acting allele may be first suspected when examination of a dataset that contains probands, mothers and fathers in a case–control study design shows a significantly increased frequency of an allele of interest among cases and mothers but not fathers compared to controls. Increased frequency in probands of an allele of interest is to be expected for a maternally acting allele because the probands will inherit alleles from their mothers simply by descent, whether an allele also acts in the probands or not. For a maternally acting allele, the expected disease allele frequencies in family members are, from highest to lowest, mothers > probands > fathers = controls [43].

The finding of increased frequency of a polymorphic allele in mothers but not fathers of affected individuals is an interesting one because it may help identify a factor that contributes to the disease. Case–control studies are powerful and highly useful but also have certain limitations. One major limitation is that the cases and controls may not be ascertained in the same way. The case–control design presumes that controls are drawn from the same population as the cases, or from a population equally at risk. Important as this presumption is, it is often not achieved in practice. For example, the allele studied may be subject to population stratification. In that situation, the case families may not be matched to the controls and the study could be invalid. Since there may be no data on population stratification for that allele, it may not be realized that the study may be inaccurate. Recent methods have ameliorated this problem when ancestrally informative markers have been genotyped [44]. Comparing parents of cases to controls, rather than parents of controls, is another example of imprecise matching [45].

If allele frequencies are significantly increased in mothers but not fathers in a case–control study, then further studies are required to determine the cause. There are several possible explanations for this pattern of increased allele frequency. The known possible reasons for increased frequency of an allele observed in mothers of affected individuals include:

(a) The allele is a maternally acting allele.
(b) The allele acts by imprinting and is imprinted in the mother.
(c) The allele acts in the affected individual and hence will have increased frequency in the parents – sometimes, by chance the allele may have increased frequency in mothers but not fathers.
(d) The allele is a mitochondrial allele and hence is transmitted only by mothers to affected individuals and therefore has increased frequency in mothers.

However, if no evidence of other mechanisms is found, then the action of a maternally acting allele is probable.

Other study designs, such as case-parent designs often used with the transmission/disequilibrium test (TDT), use internal controls, for example, by comparing the number of parental alleles transmitted to the affected child with the number of untransmitted parental alleles. Hence, population stratification is not a major limitation for TDT studies. A case–control study is an excellent first step to suggest the action of a maternally acting allele in a dataset, but such studies need to be confirmed with a more direct method.

If a case–control study has suggested that a maternally acting allele may be exerting an effect, there are at present three major designs for documenting the action of this type of suspected allele. The first is to use maternal grandparents and mothers of cases with TDT analysis for transmissions from maternal grandparents to mothers [46, 47]. Since the mother is the genetic patient for a maternally acting locus, the expectation is that such transmission disequilibrium will occur. This approach has been successful in a study of spina bifida [2] (Table 9.1, #1, 2) and two studies of autism [3, 4] (Table 9.1, #3, 4). This approach works best for disorders where the individuals with the neurodevelopmental disorder are young, since living maternal grandparents can more readily be found. This approach is direct and strongly supports the action of a maternally acting allele, but it does not address interaction between maternal and fetal genes.

A second study design uses case-parent trios analyzed with the case-parent log–linear method [48–52]. Since this method requires only the trio consisting of the individual with the neurodevelopmental disorder and the two parents, it may be the most suitable approach if living maternal grandparents are difficult to find. The test can also be extended to address the question of interaction between maternal and fetal genes [53]. With this method the data is stratified by parental mating type and the maternal genotype is further defined. For example, a mother with AA genotype and father with aa genotype represent the same mating type as a mother with aa and father with AA, but if the former pair occurs more often in parents of affected offspring, that is evidence that the AA genotype in mothers is a risk genotype for offspring. It is possible that the log–linear method may have less power than maternal TDT for the same number of families studied based upon the one report that used both methods on the same dataset [2] (Table 9.1, #1, 2).

A third test design uses "*pents*," that is, families with a child affected with a neurodevelopmental disorder, parents and maternal grandparents (five individuals per family) and analysis with a log–linear model [54]. The pent approach has the advantage of estimating both maternal and offspring genetic effects, and offers increased power, per proband, compared with the case-parent log–linear approaches. Since DNA from maternal grandparents is required, as a practical matter, it works best for early onset disorders. These three analytical approaches (case-parent design with maternal TDT analysis, case-parent design with log–linear analysis, and the pent design) are not the only possible ones, though they are the most commonly used. Other approaches have also been presented [55].

Interestingly, some of the genes identified in affected individuals by conventional approaches, especially case–control allelic association studies, and which are thought to act in affected individuals, may in fact be maternally acting alleles. This is because these affected individuals frequently receive the maternally acting alleles from their mothers simply by descent. These affected individuals will thus have an increased frequency of the maternally acting allele, as discussed earlier, despite the fact that the allele acts in the mothers, not the affected individuals themselves, to contribute the phenotype of the affected individuals. Thus, in a case–control study of affected individuals that does not include the parents, the affected individuals may have significantly increased frequency of an allele that is assumed to act in the affected individuals but in fact is a maternally acting allele. Conversely, a study comparing case-mothers to controls (or control mothers) for an allele that actually acts in the affected individual would show an increased frequency in the mothers simply because of descent.

Case Control Studies of MHC Gene Alleles and Autism

Children with autism have a change in the normal developmental pattern with impaired social interaction and communication, restricted interests, and repetitive, stereotyped patterns of behavior that can be appreciated before 36 months of age [56, 57]. Clinical genetic studies suggest that multiple gene loci interacting with each other [58, 59] and possibly with environmental and epigenetic factors [59, 60] may contribute to autism. Some of these contributing factors may be immune abnormalities.

Neuropathological studies [61, 62], cytoarchitectonic studies [63], and minicolumn studies of the cortex [64, 65] all support the prenatal origin of certain brain abnormalities in autism. Consequently, it is possible that maternal genes, acting during pregnancy, contribute to the autism phenotype in the fetus.

Beginning in 1991, Reed Warren and his group studied certain polymorphic alleles (Table 9.1, #20, 22) in the major histocompatibility complex (MHC) on chromosome 6. The MHC is discussed in more detail below. Using a case–control study design, Warren's group reported associations with autism of the null allele, *C4b*0*, of the complement gene, *C4b* (Table 9.1, #20) and of a haplotype [66] containing *C4b*0* and other MHC alleles including human leukocyte antigen-DR4 (*HLA-DR4*) (Table 9.1, #22), an allele of highly polymorphic beta chain gene *HLA-DRβ1*. He found increased frequencies of these polymorphic alleles in individuals with autism and their mothers but not in their fathers. As an explanation of their striking maternal findings, Warren et al raised the question of whether a gene acting in the mother during pregnancy might contribute to autism in her fetus [22]. Daniels et al. confirmed this finding in a subsequent case–control study [67].

Non-inherited Maternal Alleles in Rheumatoid Arthritis

In 1993, the concept of Non-inherited Maternal Alleles (NIMAs) was shown to contribute to an abnormal phenotype in rheumatoid arthritis (RA). The presence of NIMAs was suspected because the frequency of a disease allele was higher in (unaffected) mothers of patients with rheumatoid arthritis than in (unaffected) fathers. The *HLA-DR4* allele was known to contribute to RA susceptibility in patients and was shown in that study to contribute also to RA acting as a NIMA (Table 9.1, #21). This was elegantly demonstrated in families of DR4-negative RA patients where there was an increased frequency of non-inherited DR4 in the mothers of DR4-negative RA patients, compared with control families [20]. Thus, *HLA-DR4* occurring in the mother of the RA patient contributed to the patient's RA even though the patient lacked *HLA-DR4*. *HLA-DR4* occurring in the father did not have this effect. A subsequent study in 1998 [21] documented NIMA *HLA-DR4* as a susceptibility factor in two independent populations of DR4-negative RA patients. In the combined population, NIMA DR4 was significantly increased compared with non-inherited paternal allele (NIPA) DR4 (OR 3.65, 95% CI 1.29, 10.31).

The NIMA concept focuses on the non-inherited (i.e., non-transmitted) allele, in this case *HLA-DR4*. The concept is elegant because the individual affected with RA lacks *HLA-DR4* and thus, the action is likely to be a maternal effect occurring during pregnancy. However, there is no reason that a transmitted maternal allele could not have such an effect. Thus, the NIMA mechanism is useful as a proof-of-concept model for a maternal effect of *HLA-DR4*. However, at its root, NIMAs are no different from the other maternally acting alleles discussed here. The mechanism of NIMAs is essentially the same as that of maternally acting alleles.

Documentation of Maternally Acting Alleles Using More Direct Study Designs

The first report to document maternally acting alleles with a more direct study design, Doolin et al. [2], used a case-parent study design with analysis by maternal TDT (Table 9.1, #1, 2) to show that *MTR*2756G and MTRR*66G* were maternally acting alleles for spina bifida. Polymorphisms of the methionine synthase gene, *MTR* (*MTR*2756G*) and the methionine synthase reductase gene, *MTRR* (*MTRR*66G*), were associated with spina bifida in maternal trios (called mother trios), consisting of mothers of affected individuals and their parents. Maternal TDT was significant in both cases ($p = 0.004$ for *MTR*2756G* and $p = 0.05$ for *MTRR*66G*). The log–linear method was also used: these results were not statistically significant ($p = 0.10$ for *MTR*2756G* and $p = 0.08$ for *MTRR*66G*). Modeling with the log–linear approach suggested that the risk increased with the number of maternal high-risk alleles. This result was particularly interesting and important because a number of previous studies had attempted without success to associate *MTR*2756* with spina bifida using either case–control studies of individuals with spina bifida or a case TDT study design (affected individuals and parents). The rationale for suspecting this as a candidate gene was extremely strong. Maternal deficiencies of both folate and vitamin B12 had been associated with spina bifida in the fetus and MTR is the only enzyme in humans requiring both folate and vitamin B12 for its action. The presumption in the earlier negative studies was that these alleles acted in the individuals affected with spina bifida rather than in the mothers. When the possibility of a maternally acting gene allele was considered, this excellent candidate gene was indeed found to be associated with spina bifida.

Our group was the first to use a case-parent study design with analysis by maternal TDT to document a maternally acting allele for autism (Table 9.1, #3). Polymorphisms of the glutathione *S*-transferase P1 (*GSTP1*) gene (*GSTP1*313A* and *GSTP1*341C*, i.e., *GSTP1*val*^{105} and *GSTP1*ala*^{114}, respectively) were associated with autism as maternally acting alleles, using maternal TDT in trios consisting of mothers of individuals with autism and their parents [3] (Table 9.1, #3). *GSTP1*val*^{105} contributed all or most of the haplotype effect observed. GSTP1 is a major enzyme that contributes to detoxifying xenobiotics and reducing oxidative stress. Another function of GSTP1 is regulating c-Jun N-terminal kinase (JNK) by binding to JNK, in which capacity it may also regulate oxidative stress. Interestingly, *GSTP1*val*^{105} lies within the H-site of the GSTP1 protein, the region where electrophilic toxins, xenobiotics or metabolites bind to GSTP1 for conjugation with glutathione (GSH), detoxification and excretion. Also, both *GSTP1* polymorphisms lie within the region contributing to binding of the GSTP1 protein to the JUN–JNK complex.

This finding raises the question of whether oxidative stress originating in mothers could contribute to or potentiate the autism phenotype in affected fetuses.

We also used a case-parent study design with analysis by maternal TDT to document *HLA-DR4* as a maternally acting allele for autism (Table 9.1, #4). This study is discussed below.

Known Maternally Acting Alleles

At present at least 35 known reports of maternally acting alleles can be cited, most of them recently reported, and most of which play a role in causing neurodevelopmental disorders (Table 9.1, #1–35). The studies reported are of varying sample sizes, used varying analytical methods and document the presence of maternally acting alleles with proofs of varying levels of certainty. The studies themselves were discussed in detail in a recently published review [33].

Reports of *HLA-DR4* Studies in Autism

The Major Histocompatibility Complex

The MHC is a collection of immune-related genes, the largest collection in the genome. The MHC region, on chromosome 6 in humans, contains over 200 genes that have been grouped as class I, class II, and class III genes. Class I loci include *HLA-A*, *HLA-B*, and *HLA-C*. Class II loci include *HLA-DR*, *HLA-DQ*, and *HLA-DP* genes. Class I loci code for MHC class I molecules. Class I molecules are proteins that can bind stably to peptides derived from proteins synthesized and degraded in the cytosol, such as viruses. Class II loci code for MHC class II molecules. Class II molecules are proteins that can bind stably to peptides derived from proteins degraded in endocytic vesicles, such as foreign antigens or pathogens. Thus class I and class II molecules deliver peptides originating from different cell compartments to the cell surface, where their complexes are recognized by different types of T cells that can carry out immune responses. If these processes occur in the mother during pregnancy, she could recognize different environmental components as foreign and initiate inflammatory responses. Class III loci include certain complement-related genes, e.g., *C4A* and *C4B*. These genes are generally polymorphic, while some of them are highly polymorphic. The region has limited recombination in most areas except for a few recombination hot-spots. Since recombination is limited in areas of the MHC, linkage disequilibrium is seen over long distances leading to the presence of MHC-extended haplotypes. The MHC gene *HLA-DR*, and in particular its highly polymorphic beta chain gene, *HLA-DRβ1* has been studied in autism.

Case–Control Studies of HLA-DR and Autism

Warren and coauthors studied the frequency of an MHC-extended haplotype that contains *HLA-DR4* in a case–control design. This haplotype, *B44-S30-DR4*, is composed of *HLA-B44*, the *S* allele of the *BF* gene, the *three* allele of *C4A*, the *0* or *null* allele of *C4B*, and the *DR4* allele. Compared with controls, *B44-SC30-DR4* was significantly increased in both children with autism and their mothers but not their fathers [66]. Daniels et al. confirmed this finding in a subsequent case–control study [67]. Subsequently, Burger and Warren found that certain alleles of the third hypervariable region (HVR-3) of *HLA-DRβ1* had very strong association with children with autism, especially alleles within *HLA-DR4* [22] (Table 9.1, #21).

Torres et al. carried out a case–control study of HLA allele frequencies in individuals with autism spectrum disorder (ASD) compared with control Caucasian allele frequencies from the National Marrow Donor Program (NMDP) and found that *HLA-DR4* occurred more frequently in children with ASD than controls [68].

Recently, Lee et al. carried out a case–control study of *HLA-DR4* in autism in families from eastern Tennessee and families from the Autism Genetic Resource Exchange (AGRE). Tennessee families had singleton births of children with autism, while the latter group was selected from all parts of the USA and included families in which multiple males carried the diagnosis of autism. The Tennessee and AGRE families were compared to a control group of healthy unrelated adults [18] (Table 9.1, #19). Compared with controls, children with autism in the east Tennessee group and their mothers, but not their fathers, had a significantly higher frequency of *HLA-DR4* alleles than did control subjects. The mothers were 5.54 times (95% confidence interval (CI) 1.74, 18.67) and their children with autism 4.20 times (95% CI 1.37, 13.27) more likely to have *HLA-DR4* than control individuals [18]. However, *HLA-DR4* frequencies of the children with autism from the AGRE repository, their mothers and their fathers were not significantly different from controls [18].

The authors interpreted their findings in the eastern Tennessee group as consistent with a hypothesis that maternal–fetal immune interaction in utero could affect fetal brain development; such an immune interaction could conceivably involve both HLA and related genes via genetic, epigenetic, and environmental mechanisms. The authors raised the question of whether the disparate findings between the geographically defined East Tennessee group and the nationally distributed AGRE group might suggest that some environmental exposure or combinations of mechanisms were involved. They noted the stable residence of most of their east Tennessee mothers in the area. Environmental triggers such as abnormal atmospheric conditions observed there could stimulate immune activation in the mother and fetus that could contribute to the development of autism. This atmospheric pattern is associated with an unusually high incidence of inhalant allergies among both children and adults, which could cause active immune stimulation. In mothers with DR4, immune stimulation by infections, allergy, or as yet unknown environmental factors might induce formation of autoantibodies or cytokines that cross-react with antigens in the fetal brain.

The association of DR4 alleles in east Tennessee families could be attributable to the shared sequence in DRB1*0401 and *0404 that has been associated with susceptibility to rheumatoid arthritis, especially since Lee et al. [18] had observed increased frequencies of DRB1*0401 and *0404 in the east Tennessee mothers and their sons. The presence of this shared epitope might increase the affinity of binding of DR alleles for peptides that are derived from pathogens or other environmental triggers, and thus may have increased susceptibility to autoimmune pathogenesis in this specific environment.

Although these studies suggested a maternal effect of *HLA-DR4* for autism, all of them compared mothers with controls in case–control study designs. None of them carried out a more direct test such as the maternal TDT analysis in a case-parent study design that could document the presence of a maternally acting allele contributing to the autism phenotype [1, 2, 33, 46, 47].

Case-Parent Study of Autism (Table 9.1, #4)

Members of 31 families containing mothers and maternal grandparents as well as probands and fathers, recruited from New Jersey with the help of the autism support group, Autism New Jersey (previously named COSAC), were genotyped for *HLA-DR* by polymerase chain reaction amplification of genomic DNA and dot-blot analysis using sequence-specific oligonucleotide probes [69]. All probands were tested by the Autism Diagnostic Observational Schedule (ADOS-WPS) and the Autism Diagnostic Interview, Revised (ADI-R). Transmissions and non-transmissions of *HLA-DR4* were compared against the binned set of all other alleles. Using "mother of child with autism" as the affected phenotype, the standard TDT was applied to maternal trios consisting of mothers of individuals with autism and their parents. As a secondary test, maternal imprinting was studied with the method of Weinberg [51] and also the standard TDT with case trios (individual with autism and parents) [70]. Since these additional tests were secondary tests, a correction for multiple comparisons was not applied; however, even if a correction had been applied, the maternal TDT data would have remained statistically significant. At the observed allele frequencies among founders, the study had sufficient power to detect large effects, namely 80% power to detect an odds ratio of 3.6.

Clinical diagnoses by ADOS-WPS and ADI-R were autistic disorder for 30 of the probands and PDD-NOS for one. In 25 families, both parents were non-Hispanic whites; for the other six families, one parent in each was non-Hispanic white and the other was non-Hispanic Asian [35], Hispanic white [34], or non-Hispanic Black [1].

In the 31 autism families genotyped for *HLA-DRβ1*, statistically significant transmission disequilibrium for *HLA-DR4* was seen by TDT (Table 9.2, $p = 0.008$; odds ratio (OR) 4.67, 95% CI for OR 1.34–16.24) for transmissions to mothers of individuals with autism from maternal grandparents, supporting a role for *HLA-DR4* as a risk factor for autism acting in the mothers in this group of families. There were 14 copies transmitted and three untransmitted, as opposed to 35 and 46, respectively,

Table 9.2 TDT at the *DRβ1* locus for mothers of autism probands and, as a secondary test, for autism probands

Locus	Hypothesized risk allele	Transmissions	Non-transmissions	*p* Value	OR	95% CI for OR
DRβ1	*DR4*	14 (To mothers)	3	0.008	4.67	(1.34, 16.24)
DRβ1	*DR4*	12 (To cases)	9	0.39	1.33	(0.56, 3.16)
DRβ1	*DR4*	6.5 (To cases from mothers)	6.5	0.99	1.00	(0.34, 2.97)

OR odds ratio, *CI* confidence interval of OR

for other alleles. The mother of the PDD-NOS proband had an *HLA-DR4* transmission; when that family was dropped the result remained statistically significant ($p=0.01$; odds ratio (OR) 4.33, 95% CI for OR 1.23–15.21).

To examine an alternative possible explanation for the significantly increased frequency of *HLA-DR4* reported in autism mothers, that the *HLA-DR4* allele is a risk allele in the child through maternal imprinting and that the mothers are of necessity enriched for this allele, a secondary test was used in the children with autism and their parents, again using the standard TDT along with a test for maternal imprinting. Statistically significant transmission disequilibrium was not seen for transmissions from parents to individuals with autism themselves (Table 9.2, $p=0.39$; OR 1.33, 95% CI for OR 0.56–3.18), nor mothers specifically (Table 9.2, $p=0.99$, OR 1.00, 95% CI for OR 0.34–2.97). The Weinberg test for maternal imprinting was also not significant ($p=0.79$). These findings did not support action of *HLA-DR4* as a risk factor for autism acting in autism probands themselves either directly or through imprinting.

Possible Modes of Action by Which *HLA-DR4* Might Contribute to Autism

It is not known how *HLA-DR4* could act in mothers to influence brain development in their affected offspring. A number of possible mechanisms are suggested by recent work that could lead to further studies. Maternal *DR4* could contribute to a subset of autism cases by interacting with other risk alleles for autism, along with environmental factors, to perturb pathways affecting brain development and leading to autism. A possible environmental factor could be maternal infections during pregnancy (urinary tract, respiratory, or vaginal) reported previously as more common in autism mothers compared with controls [71]. These reports of maternal infections and subsequent immune activation could be analogous to the observation that *HLA-DR4* (along with *DR3*) is a risk allele for Type I diabetes mellitus and appears to modulate the humoral immune response to enterovirus antigens [72]; the patients with Type I diabetes mellitus with risk alleles DR4 and DR3 had a stronger immune response to enteroviral antigens than those with DR2. It is possible that autism mothers with *HLA-DR4* could be generating a stronger immune response to infections during pregnancy than those with other genotypes, and that this could contribute to the autism phenotype in their affected offspring through cellular effects from immune activation.

Interestingly, in mice, maternal immune activation during gestation may affect developmental outcome in their offspring. Maternal immune stimulation may be harmful, as is possible in autism, but can also be helpful. For example, maternal immune stimulation reportedly ameliorated malformations induced by chemical teratogens [73], perhaps through maternal immune regulation of fetal gene expression, including cell cycle/apoptotic genes [74]. Maternal stimulation with IFN-gamma decreased the severity of fetal cleft lip and palate caused by urethane whereas stimulation with Freund's complete adjuvant reduced both the incidence and severity of the lesion [75]. Maternal immune stimulation inducing inflammation increased fetal brain cytokine responses, decreased the number of reelin-immunoreactive cells in certain areas of postnatal brain in the offspring and altered behavior in adult offspring [76]. Reelin is a key protein in mammalian cortical development. For example, Reelin signaling influences cortical neuronal migration, which could be important for brain development in autism. Reelin gene polymorphisms have been associated with autism [77, 78], and reelin protein levels are decreased in autism in blood [79] and cerebellum [80].

Inflammation and Oxidative Stress

A number of studies support the concept that oxidative stress could play a role in causing or maintaining the symptoms of autism. It is thought that a state of elevated inflammation is part of normal pregnancy, involving both the mother and the placenta. If this inflammation is amplified during pregnancy, as from maternal infection, this might increase the risk of autism in the child [81]. The immune system can be a major source of oxidative stress during an infection, and it is possible that HLA-DR4 could act to increase maternal oxidative stress and thereby alter fetal brain development. For example, HLA-DR, as part of the immune system could "recognize" certain environmental components as foreign, and start an inflammatory process. Moreover, alterations in levels of enzymes that normally diminish oxidative stress by clearing its inducers and metabolic by-products could potentiate this effect. Polymorphic mutations in genes coding for these types of enzymes have been reportedly associated with autism and include paraoxonase [82], *GSTM1* [83], glutathione peroxidase-1 (*GPX1*) [84] and glutathione *S*-transferase-P1 (*GSTP1*) [3]. In the case of *GSTP1*, as discussed earlier, we demonstrated that the effect on autism resulted from a maternally acting allele and was thus likely to contribute to autism during pregnancy [3]. Results from a number of biochemical studies support this idea: total glutathione was observed to be significantly decreased and oxidized glutathione significantly increased in autism [85]. We found that a standard marker of oxidative stress, 8-isoprostane, was significantly increased in urine from individuals with autism [86].

Ling et al. recently demonstrated [87] that a rheumatoid arthritis (RA) shared epitope (SE) acts as a signaling ligand that activates a nitric oxide-mediated pro-oxidative pathway and blocks a cAMP-mediated antioxidative pathway, leading to increased vulnerability to oxidative damage. The SE is a product of alleles encoding certain amino acid sequences in positions 70–74 of the HLA-DRβ chain, and many of them correspond to subtypes of HLA-DR4, including the most common DR4 subtypes. This action of the SE, leading to increased vulnerability to oxidative damage, might also contribute to abnormal brain development in autism. As discussed earlier, non-inherited maternal alleles of *HLA-DR* have been shown to play a role in RA [88], indicating an effect during pregnancy that contributes to RA. Interestingly increased *HLA-DR* homozygosity both in the women with pre-eclampsia and in their partners has been strongly associated with pre-eclampsia [89], a disorder occurring in utero in which oxidative stress plays a role.

A number of possible mechanisms exists by which maternal inflammation (and thus oxidative stress) during pregnancy can affect neuronal development [90]. The neuropoietic cytokine family including IL-6, IL-11, IL-27, and others, appears to play an important role in nervous system development as well as in affecting neuronal plasticity [91]. In a rat model using the bacterial cell wall product lipopolysaccharide (LPS) to initiate maternal inflammation and mimic maternal infection during pregnancy, it was shown that cytokines mediate effects of prenatal infection on the fetus, a finding that has implications for neurodevelopmental disorders [92]. In another rodent model, the inflammatory/cytokine response from maternal injection of LPS affected both placenta and fetal brain [93]. Prenatal exposure to maternal inflammation induced by LPS in rodents, mimicking maternal infection, altered cytokine expression in placenta, amniotic fluid, and fetal brain [94]. A separate approach to inducing maternal inflammation used poly I:C (polyriboinosinic polyribocytidylic acid) exposure and found that this regulates tumor necrosis factor-alpha, brain-derived neurotrophic factor (BDNF), and nerve growth factor (NFG) expression in neonatal brain and the maternal–fetal unit of the rat [95]. These cytokines and growth factors are necessary for normal brain development and may be dysregulated due to increased maternal inflammation. Notably, elevated peripheral blood cytokine levels, which could reflect oxidative stress, have been reported in children with autism spectrum disorders [96].

Synaptic Pruning and the MHC

An interesting alternative hypothesis to the oxidative stress model proposes that HLA-DR4 plays a role in synaptic pruning, the normal process of eliminating excess connections during a child's process of learning and neural development. Synaptic pruning follows the initial overproduction of synapses during normal childhood brain development. It was recently reported that the immune system affects brain development through its role in synaptic pruning; the classical complement cascade appears to mediate elimination of synapses in the central nervous system [97]. It has been hypothesized that the immune system, by means of HLA, contributes to marking which synapses are to be eliminated, and then to pruning them by means of complement. Since this mechanism seems to involve the complement system it is not clear how HLA-DR4 might contribute to it. However, portions of the complement system, class III markers, are in linkage disequilibrium with *HLA-DR4* and thus could be involved through an MHC-extended haplotype. Still, this mechanism involves only genes acting in the fetus, whereas the *HLA-DR4* effect demonstrated thus far is in the mother.

Genes coding for proteins of the postsynaptic density (PSD) have been shown to be associated with autism, supporting the idea that synaptic abnormalities contribute to this disorder. MHC class I protein co-localizes with PSD-95 postsynaptically in dendrites and appears to contribute to the regulation of synaptic function during development, at least in experimental animals [98]. MHC class I or Ib antigens are required for regulation of synaptic pruning on neuronal bodies undergoing retrograde degeneration after axonal transection [99]. In an experimental animal model, MHC class I molecules were involved in the stripping off of synapses after an axonal lesion [100]. Again, the mechanism for a possible association of HLA-DR4, a class II molecule, with MHC class I or Ib is unclear. However, as noted, there is strong linkage disequilibrium between HLA-DR and class I markers, and these could both be involved through an MHC-extended haplotype.

HLA-DR4, High Relative Birth Weight and Possible Relevance to Autism

Autism is associated with growth abnormalities in the brain both in utero and in postnatally, i.e., the striking increase in brain growth early in postnatal life. The fetal haplotype *HLA-DR4_DQB1*0302*, the haplotype conferring the highest risk for Type I diabetes mellitus, is reportedly associated with intrauterine growth alterations, such as increased relative birth weight, in normal pregnancies [101] as well as in pregnancies of mothers with Type I diabetes [102]. This effect is aggravated by gestational infections (gestational fever, gastroenteritis or both [103]). Since a haplotype containing *HLA-DR4* has been associated with an increased intrauterine growth rate, it is possible that *HLA-DR4* could also contribute to local abnormalities of brain growth in utero.

Maternal Antibodies and Autism

Autoantibodies to human fetal brain proteins have been identified in the maternal circulation during pregnancy and may persist for years, more frequently in mothers of individuals with autism than in control mothers [104, 105]. Maternal IgG is known to cross the placenta and may also enter the fetal circulation through the second trimester of pregnancy. Certain pathogenic maternal antibodies may cause a number of neonatal autoimmune diseases [106]. It is possible that such transplacental transfer of maternal antibodies could contribute to perturbation of fetal brain development [105, 107]. Again, the mechanism of a possible contribution of HLA-DR4 to these immune-related processes in such a way as to affect the autism phenotype is not clear.

Non-inherited Maternal Allele HLA-DR4 in Rheumatoid Arthritis

The contribution of NIMA HLA-DR4 to rheumatoid arthritis (RA) was discussed earlier [20, 21]. The authors of the studies concluded that the HLA-DR4-associated genetic susceptibility to RA was due to an effect of DR4 on T-cell receptor repertoire expression in the mother of the RA patient, and that the presence of the DR4 antigen in the mother was capable of producing this effect in her children (who later developed RA), even when DR4 was not inherited by them. In autism, the same is true for *HLA-DR4;* it is an allele in the mother which contributes a biological action during pregnancy despite the fact that it is not over-transmitted to the fetus.

Future Studies of HLA Markers and Autism

A larger study is required to confirm the actions of *HLA-DR4* in mothers of individuals with autism and also to address the question of whether interaction occurs between maternal and fetal genotypes. Since linkage disequilibrium is present over large areas of the MHC, it is possible that other genes within the MHC, e.g., *HLA-DQ*, are responsible for the association observed in mothers of individuals with autism. Some disorders associated with *HLA-DR4*, such as Type I diabetes mellitus [108, 109], also have an association with an *HLA-DQ* allele, but with a higher odds ratio for certain genotypes and haplotypes. It is important to determine which gene allele or haplotype within the MHC gives the largest effect size for association with autism.

Additional studies of the action of maternal immune system genes are also needed since therapeutic approaches to immune-related disorders, e.g., autoimmune disorders, are becoming increasingly varied and effective. Since the action of *HLA-DR4* probably occurs during pregnancy, this period is also the earliest opportunity for therapy directed toward prevention and cure of autism. In utero therapeutic approaches are currently being actively developed for many disorders of pregnancy, and we should consider neurodevelopmental disorders among those that can be effectively approached during this critical period.

References

1. Johnson WG (2003) Teratogenic alleles and neurodevelopmental disorders. Bioessays 25:464–477.
2. Doolin MT, Barbaux S, McDonnell M, Hoess K, Whitehead AS, Mitchell LE (2002) Maternal genetic effects, exerted by genes involved in homocysteine remethylation, influence the risk of spina bifida. Am J Hum Genet 71(5):1222–1226.
3. Williams TA, Mars AE, Buyske SG, Stenroos ES, Wang R, Factura-Santiago MF et al (2007) Risk of autistic disorder in affected offspring of mothers with a glutathione S-transferase P1 haplotype. Arch Pediatr Adolesc Med 161(4):356–361.
4. Johnson WG, Buyske S, Mars AE, Sreenath M, Stenroos ES, Williams TA et al (2009) *HLA-DR4* as a risk allele for autism, acting in mothers of probands possibly during pregnancy. Arch Pediatr Adolesc Med 163(6):542–546.
5. Fisher RA (1944) An "incomplete" antibody in human serum. Nature 153:771–772, cited by Race RR.
6. Komrower GM, Sardharwalla IB, Coutts JM, Ingham D (1979) Management of maternal phenylketonuria: an emerging clinical problem. Br Med J 1(6175):1383–1387.
7. Hollister JM, Laing P, Mednick SA (1996) Rhesus incompatibility as a risk factor for schizophrenia in male adults. Arch Gen Psychiatry 53:19–24.
8. Palmer CG, Turunen JA, Sinsheimer JS, Minassian S, Paunio T, Lonnqvist J et al (2002) RHD maternal–fetal genotype incompatibility increases schizophrenia susceptibility. Am J Hum Genet 71(6):1312–1319.
9. Brody LC, Conley M, Cox C, Kirke PN, McKeever MP, Mills JL et al (2002) A polymorphism, R653Q, in the trifunctional enzyme methylenetetrahydrofolate dehydrogenase/methenyltetrahydrofolate cyclohydrolase/formyltetrahydrofolate synthetase is a maternal genetic risk factor for neural tube defects: report of the Birth Defects Research Group. Am J Hum Genet 71(5):1207–1215.
10. Chen D, Hu Y, Yang F, Li Z, Wu B, Fang Z et al (2005) Cytochrome P450 gene polymorphisms and risk of low birth weight. Genet Epidemiol 28(4):368–375.
11. Jensen LE, Hoess K, Mitchell LE, Whitehead AS (2006) Loss of function polymorphisms in NAT1 protect against spina bifida. Hum Genet 120(1):52–57.
12. Jensen LE, Etheredge AJ, Brown KS, Mitchell LE, Whitehead AS (2006) Maternal genotype for the monocyte chemoattractant protein 1 A(-2518)G promoter polymorphism is associated with the risk of spina bifida in offspring. Am J Med Genet A 140(10): 1114–1118.
13. Descamps OS, Bruniaux M, Guilmot PF, Tonglet R, Heller FR (2004) Lipoprotein concentrations in newborns are associated with allelic variations in their mothers. Atherosclerosis 172(2):287–298.
14. Carroll WD, Lenney W, Child F, Strange RC, Jones PW, Fryer AA (2005) Maternal glutathione S-transferase GSTP1 genotype is a specific predictor of phenotype in children with asthma. Pediatr Allergy Immunol 16(1):32–39.
15. Martinelli M, Scapoli L, Pezzetti F, Carinci F, Carinci P, Stabellini G et al (2001) C677T variant form at the MTHFR gene and CL/P: a risk factor for mothers? Am J Med Genet 98(4):357–360.
16. van Beynum IM, Kapusta L, den Heijer M, Vermeulen SH, Kouwenberg M, Daniels O et al (2006) Maternal MTHFR 677C>T is a risk factor for congenital heart defects: effect modification by periconceptional folate supplementation. Eur Heart J 27(8):981–987.
17. Rai AK, Singh S, Mehta S, Kumar A, Pandey LK, Raman R (2006) MTHFR C677T and A1298C polymorphisms are risk factors for Down's syndrome in Indian mothers. J Hum Genet 51(4):278–283.
18. Lee LC, Zachary AA, Leffell MS, Newschaffer CJ, Matteson KJ, Tyler JD et al (2006) HLA-DR4 in families with autism. Pediatr Neurol 35(5):303–307.

19. Warren RP, Singh VK, Cole P, Odell JD, Pingree CB, Warren WL et al (1991) Increased frequency of the null allele at the complement C4b locus in autism. Clin Exp Immunol 83:438–440.
20. ten Wolde S, Breedveld FC, de Vries RR, D'Amaro J, Rubenstein P, Schreuder GM et al (1993) Influence of non-inherited maternal HLA antigens on occurrence of rheumatoid arthritis. Lancet 341(8839):200–202.
21. van der Horst-Bruinsma I, Hazes JM, Schreuder GM, Radstake TR, Barrera P, van de Putte LB et al (1998) Influence of non-inherited maternal HLA-DR antigens on susceptibility to rheumatoid arthritis. Ann Rheum Dis 57(11):672–675.
22. Warren RP, Odell JD, Warren WL, Burger RA, Maciulis A, Daniels WW et al (1996) Strong association of the third hypervariable region of HLA-DR beta 1 with autism. J Neuroimmunol 67:97–102.
23. Zusterzeel PL, Nelen WL, Roelofs HM, Peters WH, Blom HJ, Steegers EA (2000) Polymorphisms in biotransformation enzymes and the risk for recurrent early pregnancy loss. Mol Hum Reprod 6(5):474–478.
24. De Marco P, Calevo MG, Moroni A, Arata L, Merello E, Cama A et al (2001) Polymorphisms in genes involved in folate metabolism as risk factors for NTDs. Eur J Pediatr Surg 11(Suppl 1):S14–S17.
25. O'Leary VB, Parle-McDermott A, Molloy AM, Kirke PN, Johnson Z, Conley M et al (2002) MTRR and MTHFR polymorphism: link to Down syndrome? Am J Med Genet 107(2):151–155.
26. van Rooij IA, Wegerif MJ, Roelofs HM, Peters WH, Kuijpers-Jagtman AM, Zielhuis GA et al (2001) Smoking, genetic polymorphisms in biotransformation enzymes, and nonsyndromic oral clefting: a gene–environment interaction. Epidemiology 12(5):502–507.
27. Gonzalez-Herrera LJ, Flores-Machado MP, Castillo-Zapata IC, Garcia-Escalante MG, Pinto-Escalante D, Gonzalez-Del Angel A (2002) Interaction of C677T and A1298C polymorphisms in the MTHFR gene in association with neural tube defects in the State of Yucatan, Mexico. Am J Hum Genet 71(4):367 (abstract).
28. Sata F, Yamada H, Kondo T, Gong Y, Tozaki S, Kobashi G et al (2003) Glutathione S-transferase M1 and T1 polymorphisms and the risk of recurrent pregnancy loss. Mol Hum Reprod 9(3):165–169.
29. Johnson WG, Stenroos ES, Spychala J, Buyske S, Chatkupt S, Ming X (2004) A new 19 bp deletion polymorphism in intron-1 of dihydrofolate reductase (DHFR) – a risk factor for spina bifida acting in mothers during pregnancy? Am J Med Genet 124A(4): 339–345.
30. Suryanarayana V, Deenadayal M, Singh L (2004) Association of CYP1A1 gene polymorphism with recurrent pregnancy loss in the South Indian population. Hum Reprod 19(11):2648–2652.
31. Johnson WG, Scholl TO, Spychala JR, Buyske S, Stenroos ES, Chen X (2005) Common dihydrofolate reductase 19 bp deletion allele: a novel risk factor for preterm delivery. Am J Clin Nutr 81:664–668.
32. Scala I, Granese B, Sellitto M, Salome S, Sammartino A, Pepe A et al (2006) Analysis of seven maternal polymorphisms of genes involved in homocysteine/folate metabolism and risk of Down syndrome offspring. Genet Med 8(7):409–416.
33. Johnson WG, Sreenath M, Buyske S, Stenroos ES (2008) Teratogenic alleles in autism and other neurodevelopmental disorders. In: Zimmerman A (ed) Autism: current theories and evidence. Humana, Totowa, pp 41–68.
34. Levine P, Stetson RE (1939) An unusual case of intragroup agglutination. JAMA 113:126–127.
35. Levine P, Burnham L, Katzin EM, Vogel P (1941) The role of iso-immunization in the pathogenesis of erythroblastosis fetalis. Am J Obstet Gynecol 42:925–937.
36. Flegel WA (2006) Molecular genetics of RH and its clinical application. Transfus Clin Biol 13(1–2):4–12.

37. Yan L, Wu J, Zhu F, Hong X, Xu X (2007) Molecular basis of D variants in Chinese persons. Transfusion 47(3):471–477.
38. Westgren M, Ek S, Remberger M, Ringden O, Stangenberg M (1995) Cytokines in fetal blood and amniotic fluid in Rh-immunized pregnancies. Obstet Gynecol 86(2):209–213.
39. Guttler F, Azen C, Guldberg P, Romstad A, Hanley WB, Levy HL et al (1999) Relationship among genotype, biochemical phenotype, and cognitive performance in females with phenylalanine hydroxylase deficiency: report from the Maternal Phenylketonuria Collaborative Study. Pediatrics 104(2 Pt 1):258–262.
40. Koch R, Levy HL, Matalon R, Rouse B, Hanley WB, Trefz F et al (1994) The international collaborative study of maternal phenylketonuria: status report 1994. Acta Paediatr Suppl 407:111–119.
41. Allen RJ, Brunberg J, Schwartz E, Schaefer AM, Jackson G (1994) MRI characterization of cerebral dysgenesis in maternal PKU. Acta Paediatr Suppl 407:83–85.
42. Menkes JH (1990) Textbook of child neurology, 4th edn. Lea & Febiger, Philadelphia.
43. Buyske S (2008) Maternal genotype effects can alias case genotype effects in case-control studies. Eur J Hum Genet 16(7):783–785.
44. Tiwari HK, Barnholtz-Sloan J, Wineinger N, Padilla MA, Vaughan LK, Allison DB (2008) Review and evaluation of methods correcting for population stratification with a focus on underlying statistical principles. Hum Hered 66(2):67–86.
45. Shi M, Umbach DM, Vermeulen SH, Weinberg CR (2008) Making the most of case-mother/control-mother studies. Am J Epidemiol 168(5):541–547.
46. Mitchell LE (1997) Differentiating between fetal and maternal genotypic effects, using the transmission test for linkage disequilibrium. Am J Hum Genet 60:1006–1007.
47. Johnson WG (1999) The DNA polymorphism-diet-cofactor-development hypothesis and the gene-teratogen model for schizophrenia and other developmental disorders. Am J Med Genet (Neuropsychiatr Genet) 88:311–323.
48. Weinberg CR, Wilcox AJ, Lie RT (1998) A log–linear approach to case-parent-triad data: assessing effects of disease genes that act either directly or through maternal effects and that may be subject to parental imprinting. Am J Hum Genet 62(4):969–978.
49. Wilcox AJ, Weinberg CR, Lie RT (1998) Distinguishing the effects of maternal and offspring genes through studies of "case-parent triads". Am J Epidemiol 148(9):893–901.
50. Weinberg CR, Wilcox AJ (1999) Re: "Distinguishing the effects of maternal and offspring genes through studies of 'case-parent triads'" and "a new method for estimating the risk ratio in studies using case-parental control design". Am J Epidemiol 150(4):428–429.
51. Weinberg CR (1999) Methods for detection of parent-of-origin effects in genetic studies of case-parents triads. Am J Hum Genet 65(1):229–235.
52. Starr JR, Hsu L, Schwartz SM (2005) Assessing maternal genetic associations: a comparison of the log–linear approach to case-parent triad data and a case-control approach. Epidemiology 16(3):294–303.
53. Minassian SL, Palmer CG, Sinsheimer JS (2005) An exact maternal–fetal genotype incompatibility (MFG) test. Genet Epidemiol 28(1):83–95.
54. Mitchell LE, Weinberg CR (2005) Evaluation of offspring and maternal genetic effects on disease risk using a family-based approach: the "pent" design. Am J Epidemiol 162(7):676–685.
55. Mitchell LE, Starr JR, Weinberg CR, Sinsheimer JS, Mitchell LE, Murray JC (2005) Maternal genetic effects. Presented at Annual meeting, Concurrent invited sessions I, #14, American Society of Human Genetics, Salt Lake City UT, Wed, Oct 26, 8–10 pm, 2005, Moderator, Laura E Mitchell.
56. American Psychiatric Association (1994) Diagnostic and statistical manual of mental disorders, 4th edn. American Psychiatric Association, Washington, D.C.
57. Rapin I (1997) Autism. N Engl J Med 337:97–104.
58. Muhle R, Trentacoste SV, Rapin I (2004) The genetics of autism. Pediatrics 113(5):e472–e486.

59. Szatmari P (2003) The causes of autism spectrum disorders. Br Med J 326(7382):173–174.
60. Lawler CP, Croen LA, Grether JK, Van de Water J (2004) Identifying environmental contributions to autism: provocative clues and false leads. Ment Retard Dev Disabil Res Rev 10(4):292–302.
61. Rodier PM, Ingram JL, Tisdale B, Nelson S, Romano J (1996) Embryological origin for autism: developmental anomalies of the cranial nerve motor nuclei. J Comp Neurol 370:247–261.
62. Stromland K, Nordin V, Miller M, Akerstrom B, Gillberg C (1994) Autism in thalidomide embryopathy: a population study. Dev Med Child Neurol 36:351–356.
63. Piven J, O'Leary D (1999) Neuroimaging in autism. Child Adolesc Psychiatr Clin N Am 6:305–323.
64. Casanova MF, Buxhoeveden D, Gomez J (2003) Disruption in the inhibitory architecture of the cell minicolumn: implications for autism. Neuroscientist 9(6):496–507.
65. Casanova MF, Buxhoeveden DP, Switala AE, Roy E (2002) Minicolumnar pathology in autism. Neurology 58(3):428–432.
66. Warren RP, Singh VK, Cole P, Odell JD, Pingree CB, Warren WL et al (1992) Possible association of the extended MHC haplotype B44-SC30-DR4 with autism. Immunogenetics 36:203–207.
67. Daniels WW, Warren RP, Odell JD, Maciulis A, Burger RA, Warren WL et al (1995) Increased frequency of the extended or ancestral haplotype B44- SC30-DR4 in autism. Neuropsychobiology 32:120–123.
68. Torres AR, Maciulis A, Stubbs EG, Cutler A, Odell D (2002) The transmission disequilibrium test suggests that HLA-DR4 and DR13 are linked to autism spectrum disorder. Hum Immunol 63(4):311–316.
69. (1990) HLA typing. In: Zachary AA, Teresi GA (eds). ASHI laboratory manual, American Society for Histocompatibility and Immunogenetics, New York, p 195.
70. Spielman RS, McGinnis RE, Ewens WJ (1993) Transmission test for linkage disequilibrium: the insulin gene region and insulin-dependent diabetes mellitus (IDDM). Am J Hum Genet 52:506–516.
71. Comi AM, Zimmerman AW, Frye VH, Law PA, Peeden JN (1999) Familial clustering of autoimmune disorders and evaluation of medical risk factors in autism. J Child Neurol 14:388–394.
72. Sadeharju K, Knip M, Hiltunen M, Akerblom HK, Hyoty H (2003) The HLA-DR phenotype modulates the humoral immune response to enterovirus antigens. Diabetologia 46(8):1100–1105.
73. Holladay SD, Sharova LV, Punareewattana K, Hrubec TC, Gogal RM Jr, Prater MR et al (2002) Maternal immune stimulation in mice decreases fetal malformations caused by teratogens. Int Immunopharmacol 2(2–3):325–332.
74. Sharova L, Sura P, Smith BJ, Gogal RM Jr, Sharov AA, Ward DL et al (2000) Nonspecific stimulation of the maternal immune system. II. Effects on gene expression in the fetus. Teratology 62(6):420–428.
75. Holladay SD, Sharova L, Smith BJ, Gogal RM Jr, Ward DL, Blaylock BL (2000) Nonspecific stimulation of the maternal immune system. I. Effects On teratogen-induced fetal malformations. Teratology 62(6):413–419.
76. Meyer U, Nyffeler M, Engler A, Urwyler A, Schedlowski M, Knuesel I et al (2006) The time of prenatal immune challenge determines the specificity of inflammation-mediated brain and behavioral pathology. J Neurosci 26(18):4752–4762.
77. Serajee FJ, Zhong H, Mahbubul Huq AH (2006) Association of Reelin gene polymorphisms with autism. Genomics 87(1):75–83.
78. Skaar DA, Shao Y, Haines JL, Stenger JE, Jaworski J, Martin ER et al (2005) Analysis of the RELN gene as a genetic risk factor for autism. Mol Psychiatry 10(6):563–571.
79. Fatemi SH, Stary JM, Egan EA (2002) Reduced blood levels of reelin as a vulnerability factor in pathophysiology of autistic disorder. Cell Mol Neurobiol 22(2):139–152.

80. Fatemi SH, Stary JM, Halt AR, Realmuto GR (2001) Dysregulation of reelin and Bcl-2 proteins in autistic cerebellum. J Autism Dev Disord 31(6):529–535.
81. Patterson PH, Xu W, Smith SEP, Devarman BE (2008) Maternal immune activation, cytokines and autism. In: Zimmerman AW (ed) Autism. Current theories and evidence. Humana, Totowa, NJ, pp 289–307.
82. Serajee FJ, Nabi R, Zhong H, Huq M (2004) Polymorphisms in xenobiotic metabolism genes and autism. J Child Neurol 19(6):413–417.
83. Buyske S, Williams TA, Mars AE, Stenroos ES, Ming SX, Wang R et al (2006) Analysis of case-parent trios at a locus with a deletion allele: association of GSTM1 with autism. BMC Genet 7(1):8.
84. Ming X, Johnson WG, Stenroos ES, Mars A, Lambert GH, Buyske S (2010) Genetic variant of glutathione peroxidase 1 in autism. Brain Dev 32(2):105–109.
85. James SJ, Cutler P, Melnyk S, Jernigan S, Janak L, Gaylor DW et al (2004) Metabolic biomarkers of increased oxidative stress and impaired methylation capacity in children with autism. Am J Clin Nutr 80(6):1611–1617.
86. Ming X, Stein TP, Brimacombe M, Johnson WG, Lambert GH, Wagner GC (2005) Increased excretion of a lipid peroxidation biomarker in autism. Prostaglandins Leukot Essent Fatty Acids 73(5):379–384.
87. Ling S, Li Z, Borschukova O, Xiao L, Pumpens P, Holoshitz J (2007) The rheumatoid arthritis shared epitope increases cellular susceptibility to oxidative stress by antagonizing an adenosine-mediated anti-oxidative pathway. Arthritis Res Ther 9(1):R5.
88. Feitsma AL, Worthington J, van der Helm-van Mil AH, Plant D, Thomson W, Ursum J et al (2007) Protective effect of noninherited maternal HLA-DR antigens on rheumatoid arthritis development. Proc Natl Acad Sci U S A 104(50):19966–19970.
89. de LB I, Battini L, Simonelli M, Clemente F, Brunori E, Mariotti ML et al (2000) Increased HLA-DR homozygosity associated with pre-eclampsia. Hum Reprod 15(8):1807–1812.
90. Jonakait GM (2007) The effects of maternal inflammation on neuronal development: possible mechanisms. Int J Dev Neurosci 25(7):415–425.
91. Bauer S, Kerr BJ, Patterson PH (2007) The neuropoietic cytokine family in development, plasticity, disease and injury. Nat Rev Neurosci 8(3):221–232.
92. Ashdown H, Dumont Y, Ng M, Poole S, Boksa P, Luheshi GN (2006) The role of cytokines in mediating effects of prenatal infection on the fetus: implications for schizophrenia. Mol Psychiatry 11(1):47–55.
93. Bell MJ, Hallenbeck JM, Gallo V (2004) Determining the fetal inflammatory response in an experimental model of intrauterine inflammation in rats. Pediatr Res 56(4):541–546.
94. Urakubo A, Jarskog LF, Lieberman JA, Gilmore JH (2001) Prenatal exposure to maternal infection alters cytokine expression in the placenta, amniotic fluid, and fetal brain. Schizophr Res 47(1):27–36.
95. Gilmore JH, Jarskog LF, Vadlamudi S (2005) Maternal poly I:C exposure during pregnancy regulates TNF alpha, BDNF, and NGF expression in neonatal brain and the maternal-fetal unit of the rat. J Neuroimmunol 159(1–2):106–112.
96. Molloy CA, Morrow AL, Meinzen-Derr J, Schleifer K, Dienger K, Manning-Court P et al (2006) Elevated cytokine levels in children with autism spectrum disorder. J Neuroimmunol 172(1–2):198–205.
97. Stevens B, Allen NJ, Vazquez LE, Howell GR, Christopherson KS, Nouri N et al (2007) The classical complement cascade mediates CNS synapse elimination. Cell 131(6):1164–1178.
98. Goddard CA, Butts DA, Shatz CJ (2007) Regulation of CNS synapses by neuronal MHC class I. Proc Natl Acad Sci U S A 104(16):6828–6833.
99. Wekerle H (2005) Planting and pruning in the brain: MHC antigens involved in synaptic plasticity? Proc Natl Acad Sci U S A 102(1):3–4.
100. Thams S, Oliveira A, Cullheim S (2008) MHC class I expression and synaptic plasticity after nerve lesion. Brain Res Rev 57(1):265–269.

101. Larsson HE, Lynch K, Lernmark B, Nilsson A, Hansson G, Almgren P et al (2005) Diabetes-associated HLA genotypes affect birthweight in the general population. Diabetologia 48(8):1484–1491.
102. Hummel M, Marienfeld S, Huppmann M, Knopff A, Voigt M, Bonifacio E et al (2007) Fetal growth is increased by maternal type 1 diabetes and HLA DR4-related gene interactions. Diabetologia 50(4):850–858.
103. Larsson HE, Lynch K, Lernmark B, Hansson G, Lernmark A, Ivarsson SA (2007) Relationship between increased relative birthweight and infections during pregnancy in children with a high-risk diabetes HLA genotype. Diabetologia 50(6):1161–1169.
104. Braunschweig D, Ashwood P, Krakowiak P, Hertz-Picciotto I, Hansen R, Croen LA et al (2008) Autism: maternally derived antibodies specific for fetal brain proteins. Neurotoxicology 29(2):226–231.
105. Morris CM, Pletnikov M, Zimmerman AW, Singer HS (2008) Maternal antibodies and the placental–fetal IgG transfer theory. In: Zimmerman AW (ed) Autism. Current theories and evidence. Humana, Totowa, pp 309–328.
106. Heuer L, Ashwood P, Van de Water J (2008) The immune system in autism. Is there a connection? In: Zimmerman AW (ed) Autism. Current theories and evidence. Humana, Totowa, NJ, pp 271–288.
107. Zimmerman AW, Connors SL, Matteson KJ, Lee LC, Singer HS, Castaneda JA et al (2007) Maternal antibrain antibodies in autism. Brain Behav Immun 21(3):351–357.
108. Saruhan-Direskeneli G, Uyar FA, Bas F, Gunoz H, Bundak R, Saka N et al (2000) HLA-DR and -DQ associations with insulin-dependent diabetes mellitus in a population of Turkey. Hum Immunol 61(3):296–302.
109. Noble JA, Valdes AM, Cook M, Klitz W, Thomson G, Erlich HA (1996) The role of HLA class II genes in insulin-dependent diabetes mellitus: molecular analysis of 180 Caucasian, multiplex families. Am J Hum Genet 59(5):1134–1148.

Index

MIX
Papier aus verantwortungsvollen Quellen
Paper from responsible sources
FSC® C105338

If you have any concerns about our products,
you can contact us on
ProductSafety@springernature.com

In case Publisher is established outside the EU,
the EU authorized representative is:
Springer Nature Customer Service Center GmbH
Europaplatz 3, 69115 Heidelberg, Germany

Printed by Libri Plureos GmbH
in Hamburg, Germany